VISIBLE-LIGHT PHOTOCATALYSIS OF CARBON-BASED MATERIALS

Edited by **Yunjin Yao**

Visible-Light Photocatalysis of Carbon-Based Materials
http://dx.doi.org/10.5772/intechopen.68233
Edited by Yunjin Yao

Contributors

Ozlem Altintas Yildirim, Teoman Ozturk, Hideto Miyabe, Zhigang Zhao, Shan Cong, Haining Tian, Palas Baran Pati, Su-Il In, Masahide Yasuda, Jin Matsumoto, Abdullahi B. Makama, Fariborz Taghipour, Babak Adeli, Diana Rakhmawaty Eddy, Atiek Rostika Noviyanti, Solihudin Solihudin, Safri Ishmayana, Roekmi-Ati Tjokronegoro

Notice

First published in London, United Kingdom, 2018 by IntechOpen
IntechOpen is the global imprint of INTECHOPEN LIMITED, registered in England and Wales, registration number: 11086078, The Shard, 25th floor, 32 London Bridge Street
London, SE19SG – United Kingdom
Printed in Croatia

British Library Cataloguing-in-Publication Data
A catalogue record for this book is available from the British Library

Additional hard copies can be obtained from orders@intechopen.com

Visible-Light Photocatalysis of Carbon-Based Materials, Edited by Yunjin Yao
p. cm.
Print ISBN 978-1-78923-028-4
Online ISBN 978-1-78923-029-1

For EU product safety concerns:
IN TECH d.o.o., Prolaz Marije Krucifikse Kozulić 3, 51000 Rijeka, Croatia,
info@intechopen.com or visit our website at intechopen.com.

Meet the editor

Dr. Yunjin Yao's research focuses on synthesis, characterization, and application of functional nanomaterials and pollution control principles and applications of various environmental technologies, particularly various advanced oxidation processes in water and air. He has published more than 50 original journal publications (including top-ranking journals, such as *Applied Catalysis B: Environmental, Journal of Hazardous Materials, Water Research,* and *Chemical Engineering Journal*) and 13 patents and delivered over 20 plenary/keynote/invited presentations. His current projects include chemical and photocatalytic degradation kinetics and mechanisms of organic pollutants. Apart from his personal research, he also serves as an editorial board member of *American Chemical Science Journal, Journal of Applied Chemical Science International,* and *Journal of Analytical, Bioanalytical and Separation Techniques*.

Contents

Preface

The increasing worldwide energy shortage and environmental pollution issues are becoming serious threats to the development of human society. Over the past two decades, sunlight has been served as the most ideal power to drive the chemical reactions, resolve the energy shortage, and remove pollution. Particularly, semiconductor-based photocatalysis has been considered as an economic, safe, renewable, and clean technology, for various applications, such as photodegradation of organic compounds, water splitting, and selective organic transformations. Nevertheless, the wide band gap and the low solar-energy utilization efficiency still remain the "bottleneck" to satisfy the requirements of practical applications. Thus, a key challenge is to develop novel photocatalysts that are stable, abundant, efficient, and facile in fabrication.

Recently, highly graphitic carbons and nanostructured sp^2 carbons (e.g., nanotubes, graphenes, fibers, graphite powder, and graphitic carbon nitride (g-C_3N_4)) with excellent properties have been widely used in optical and electronic devices, chemical sensors, energy generation and storage, as well as environmental remediation. As is known, the photocatalytic process involves the generation of charge carriers, that is, electrons and holes induced by light. However, the pristine carbon material is usually restricted by unsatisfactory photocatalytic efficiency and practical applications due to the insufficient solar light absorption, the low-surface area, and the fast recombination of photogenerated electron-hole pairs. Generally, each photocatalytic reaction would involve three processes: photon absorption, electron-hole charge generation and charge separation, and catalytic surface reactions. Therefore, various modification strategies, such as elemental and molecular doping, preparation of mesoporous carbon materials, and combination of conductive materials, are adopted to enhance the photocatalytic activity of carbon materials.

In this book, we intend to describe the great potential of efficient and low-cost carbon-based materials in various realms, such as the decomposition of pollutants, the water-gas shift reaction, and the photocatalytic evolution of H_2. Finally, the current challenges and the crucial issues of carbon-based photocatalysts that need to be addressed for the future applications are presented.

Yunjin Yao
Hefei University of Technology, China

Chapter 1

Organic Reactions Promoted by Metal-Free Organic Dyes Under Visible Light Irradiation

Hideto Miyabe

Additional information is available at the end of the chapter

http://dx.doi.org/10.5772/intechopen.70507

Abstract

Although dyes have received much attention as the visible light-activated photocatalysts, the use of metal-free organic dyes in synthetic organic chemistry is still limited. This chapter summarizes the recent progress in the visible light photocatalysis promoted by metal-free organic dyes. Eosin Y is the typical organic dyes to induce the photoredox catalysis. Recently, other organic dyes such as Rose Bengal, fluorescein, and methylene blue have been studied as photocatalysts to promote the single-electron transfer processes.

Keywords: photocatalyst, organocatalyst, dye, catalysis, visible light, radical

1. Introduction

The use of abundant sunlight as a clean source of energy is an important aim of green chemistry. In recent years, dyes have attracted a great deal of attention as the visible light-activated photocatalysts in synthetic organic chemistry. However, these studies have mainly concentrated on the redox transformations using transition metal dyes such as ruthenium or iridium photocatalysts [1–9]. In contrast, the use of metal-free dyes still remains rather underdeveloped, although organic dyes are more environmentally friendly and cheaper. Eosin Y is the typical organic dyes to induce the photoredox catalysis [10]. Recently, Rose Bengal, fluorescein, methylene blue, and other organic dyes have been studied as photocatalysts to promote the single-electron transfer processes [11–13]. Additionally, 3-cyano-1-methylquinolinium, 9-mesityl-10-methylacridinium ion, and acridinium salts were developed as organic photocatalysts [12, 13].

The photoredox cycle is initiated by the visible light irradiation of dye in the ground state to produce the high-energy excited state of dye (Dye*) (**Figure 1**). Two distinctive pathways

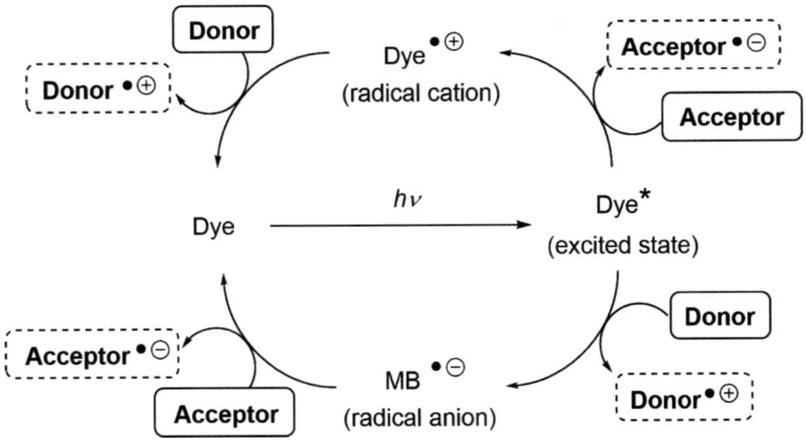

Figure 1. Dye-catalyzed photoredox cycle.

from dye in the excited state (Dye*) are described for the mechanism of visible light photoredox catalysis. The reductive property of Dye* can be used in the presence of a sacrificial electron acceptor. In other words, Dye* serves as an electron donor leading the radical cation species of Dye. In contrast, Dye* also acts as an electron acceptor in the presence of a sacrificial electron donor.

2. Eosin Y and eosin B

Eosin Y (EY) is the typical organic dye to induce the synthetically useful photoredox transformations [14–30]. EY that absorbed visible light populates in the lowest excited singlet state. The subsequent spin-forbidden singlet-triplet intersystem crossing affords EY in the excited triplet state. A variety of photoredox transformations are induced by the single-electron transfer from EY in the excited triplet state. In 2014, König provides the review article concerning the utility of EY as a photocatalyst in synthetic organic chemistry [10]. Therefore, this section highlights the recent remarkable progress in the EY-catalyzed photoredox transformations.

The EY-catalyzed generation of aryl radicals from aryl diazonium salts was studied by König's group [14]. The direct C–H bond arylation of heteroarenes with aryl diazonium salts was achieved by employing only 1 mol% of EY (**Figure 2**). The arylation of furan with diazonium salt **1** in dimethyl sulfoxide (DMSO) proceeded smoothly under the visible light irradiation to give the desired coupling product **2** in 85% yield. This transformation proceeds through the radical mechanism. Initially, aryl radical is produced by the single-electron transfer from the excited EY (EY*) to aryl diazonium salt **1**. The addition of aryl radical to furan leads to the formation of the radical intermediate **A**, which is further oxidized to cation intermediate **B**. Final deprotonation gives the coupling product **2**. Next, the EY-catalyzed arylation of simple arenes with fluorinated aryl bromides was developed [15]. In the presence of EY (5 mol%) and triethylamine as an electron donor, the direct arylation using 1-bromo-2,3,4,5,6-pentafluorobenzene **3** and benzene gave the coupling product **4** in 85% yield. The mechanistic investigations reveal that the photooxidation of

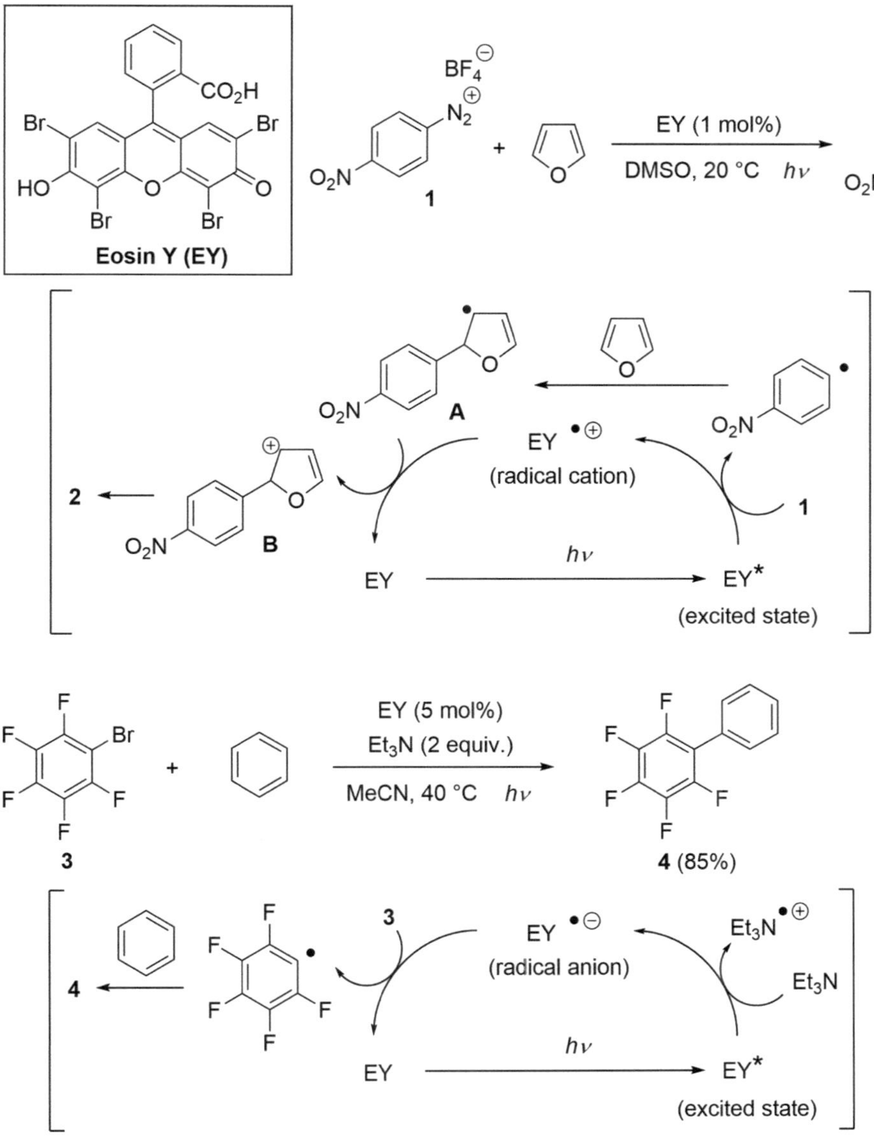

Figure 2. EY-catalyzed generation of aryl radicals and coupling reactions.

triethylamine by the excited EY (EY*), and the subsequent single-electron transfer from the radical anion species of EY to **3** leads to the formation of the polyfluorinated aryl radical. The mild visible light-mediated generation of aryl radicals from diazonium salts was also investigated by Wangelin's group [16–18]. The coupling reaction catalyzed by EY was investigated with no use of any sacrificial oxidants [19, 20].

The vinyl sulfones were synthesized by the EY-catalyzed reaction of alkenes with sodium aryl sulfinates [21]. The reaction of 1,2-dihydronaphthalene **6** with sodium benzenesulfinate **5** was performed in the presence of EY (10 mol%) and nitrobenzene as a terminal oxidant (**Figure 3**).

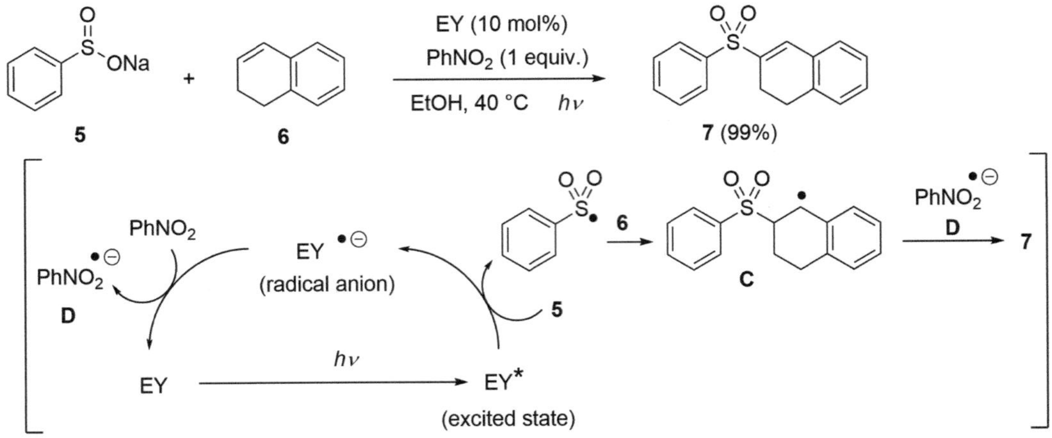

Figure 3. EY-catalyzed reaction using sodium benzenesulfinate.

The desired vinyl sulfone **7** was obtained in 99% yield. In this reaction, sodium sulfinate **5** is oxidized by the excited EY (EY*) to give the sulfonyl radical, which attacks the double bond of **6** to form the radical intermediate **C**. Nitrobenzene oxidizes the radical cation species of EY to give EY in the ground state and the radical anion species **D** of nitrobenzene, which reacts with radical intermediate **C** to give the vinyl sulfone **7**.

The oxidative cyclization reaction between 3-phenylpropiolate **8** and 4-methylbenzenesulfinic acid **9** was studied (**Figure 4**) [22]. In the presence of EY (1 mol%) and *tert*-butyl hydroperoxide (TBHP), the reaction between **8** and **9** was performed in MeCN–H_2O (1:1, v/v) under the

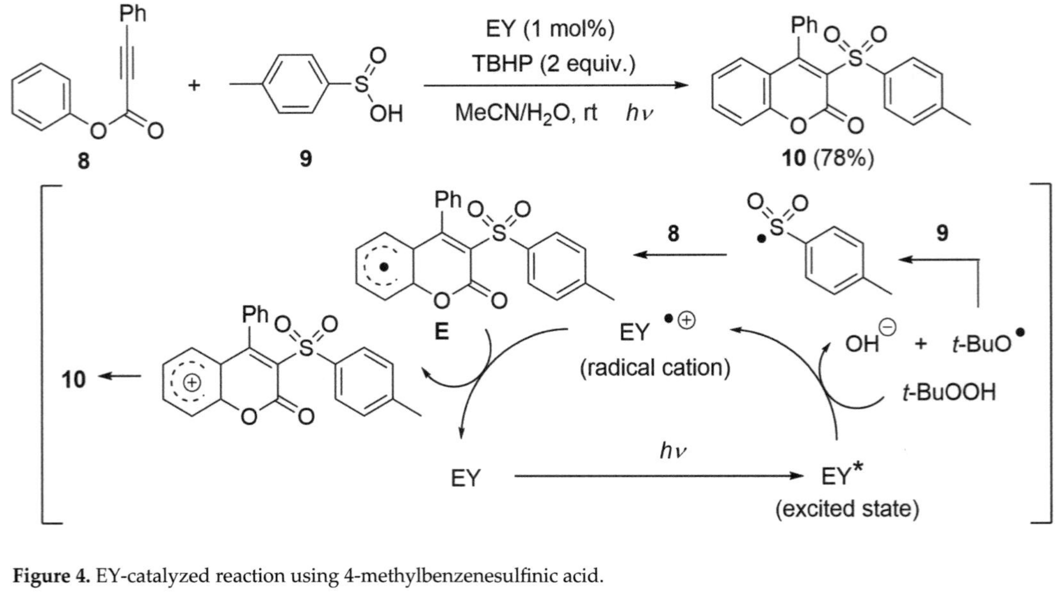

Figure 4. EY-catalyzed reaction using 4-methylbenzenesulfinic acid.

visible light irradiation. The desired coumarin **10** was obtained in 78% isolated yield. Initially, *tert*-butoxyl radical is produced by the single-electron transfer from the excited EY (EY*) to TBHP. The cyclization reaction is promoted by the addition of sulfonyl radical, generated from sulfinic acid **9** and *tert*-butoxyl radical, to alkyne moiety of **8**. The coumarin **10** is formed *via* the oxidation of the cyclized radical intermediate **E** by the radical cation species of EY.

The EY-catalyzed cyclization of 2-isocyanobiphenyls with arylsulfonyl chlorides took place under the oxidant-free visible light irradiation conditions [23]. In the presence of K_2HPO_4 as a base, the EY-catalyzed reaction of benzenesulfonyl chloride **11** and 2-isocyanobiphenyl **12** proceeded smoothly to give the 6-phenyl-substituted phenanthridine **13** in 79% yield (**Figure 5**). Initially, the single-electron transfer from the excited EY (EY*) to sulfonyl chloride **11** gives the phenyl radical, which adds to isocyanide **12** to form the imidoyl radical intermediate **F**. The subsequent cyclization gives the cyclized radical intermediate **G**, which is oxidized by the radical cation species of EY. Finally, the deprotonation leads to **13**.

EY could be used as the photocatalyst for the 5-*exo*-trig cyclization of iminyl radicals generated from *O*-aryl oximes [24]. Among several aryl oximes evaluated, 2,4-dinitro-substituted aryl oxime **14** has the excellent reactivity due to its low reduction potential (**Figure 6**). In the presence of cyclohexadiene (CHD) as a H-donor, EY-catalyzed photoreaction of **14** gave the cyclized product **15** in 78% yield. In this transformation, the iminyl radical **I** is generated *via* the reduction of **14** by the excited EY (EY*) followed by the fragmentation of radical anion **H**. The cyclization of **I** gives the cyclized *C*-centered radical **J**, which abstracts H-atom from CHD to give the desired product **15**. Furthermore, the formation of product **15** was observed even in the absence of EY, when the MeCN solution of **14** was treated with Et_3N under the visible

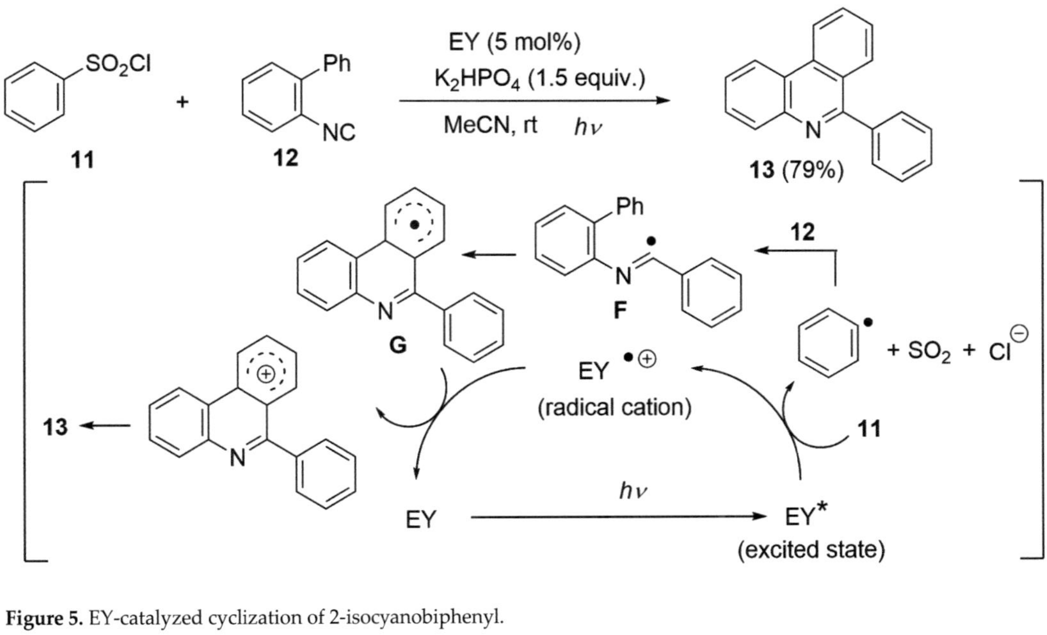

Figure 5. EY-catalyzed cyclization of 2-isocyanobiphenyl.

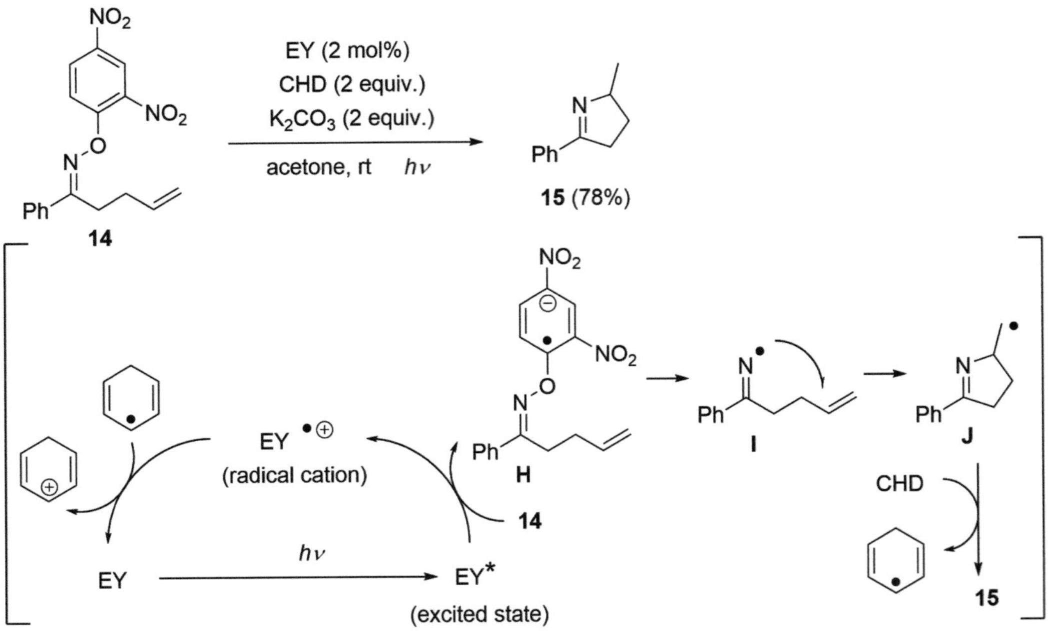

Figure 6. EY-catalyzed cyclization of 2,4-dinitro-substituted aryl oxime.

light irradiation. In this case, the visible light-mediated electron transfer would be induced by the formation of donor-acceptor complex between Et_3N and 2,4-dinitrophenyl group of **14**.

The EY-induced photocatalysis was applied to the radical cascade cyclization of polyenes [25]. The photocatalytic cascade cyclization of polyene **16** proceeded by employing EY (**Figure 7**). Hexafluoro-2-propanol (HFIP) was identified as the optimal solvent. The cyclized product **17** was obtained in 93% yield with the excellent diastereoselectivity *via* the radical cation intermediate **K** generated by the single-electron transfer from **16** to the excited EY (EY*). In this process, the OH moiety of **16** would act as a terminator.

Eosin B is also the active catalyst under the visible light irradiation [31]. The C–H functionalization of thiazole derivatives with diarylphosphine oxides was achieved by the eosin B-catalyzed photoredox process. When eosin B was employed as a photocatalyst, the phosphorylation of benzothiazole **18** with diphenylphosphine oxide proceeded effectively to give the phosphorylation product **19** in 87% yield (**Figure 8**). In this transformation, hydrogen (H_2) is the only by-product.

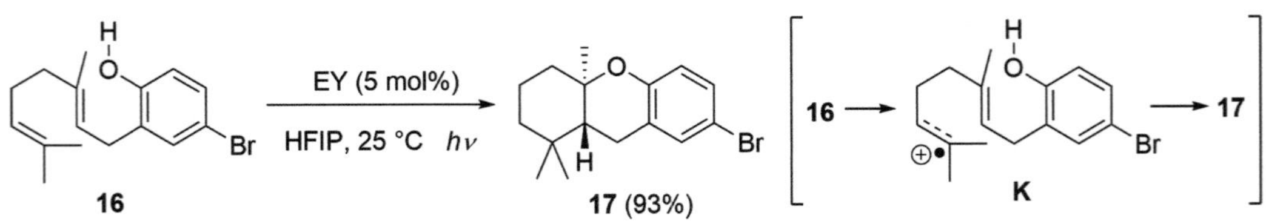

Figure 7. EY-catalyzed cascade cyclization of polyene.

http://dx.doi.org/10.5772/intechopen.70507

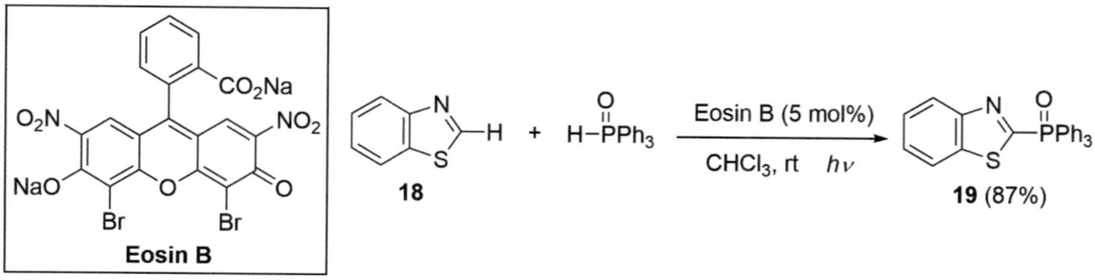

Figure 8. Eosin B-catalyzed phosphorylation of benzothiazole.

3. Rose Bengal

Rose Bengal (RB) was widely used as a visible light-activated photocatalyst [32–44]. Tan's group studied the photoredox catalysis using RB [32–36]. RB was a good catalyst for the dehydrogenative coupling reaction between tetrahydroisoquinolines and nitroalkanes (**Figure 9**) [33]. In the presence of RB (5 mol%), the reaction of *N*-phenyl-tetrahydroisoquinoline **20** with nitromethane gave the adduct **21** in 92% yield. In the absence of O_2, a much lower yield was obtained relative to the reaction performed in open air; thus, air is important for this reaction as an oxidant. Additionally, this reaction was expanded to the dehydrogenative Mannich reaction using enamine nucleophiles generated from ketones and pyrrolidine.

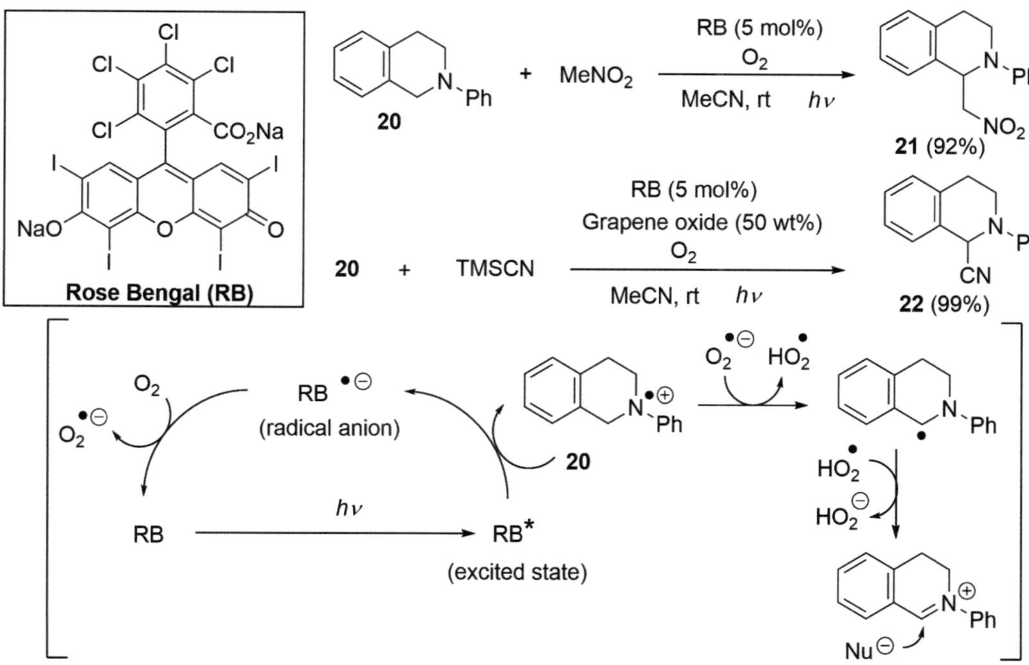

Figure 9. RB-catalyzed reaction of tetrahydroisoquinoline.

Next, the combination of graphene oxide and RB was studied in the reaction between tetrahydroisoquinolines with TMSCN or $TMSCF_3$ [34]. In the presence of RB (5 mol%) and graphene oxide (50 wt%), the reaction of **20** with TMSCN proceeded effectively to give the adduct **22** in 99% yield, while the yield of **22** decreased to 45% in the absence of graphene oxide. The use of graphene oxide as a cocatalyst improves the reaction rates and yields.

New method for the synthesis of Meyers's bicyclic lactams was developed by using RB photocatalysis [37–39]. This cascade transformation is the one-pot reaction which begins from furan substrates (**Figure 10**) [37]. Despite the extraordinary complexity of reaction cascade, the reaction between 2-methylfuran **23** and L-serine ethyl ester **24** led to the formation of bicyclic lactam **25** in 68% yield. At first, RB promotes the photooxidation of methylfuran **23** with singlet oxygen in MeOH. The intermediate **L** is formed by the in situ reduction of hydroperoxy with Me_2S. The next reaction of **L** with L-serine ethyl ester **24** gives the intermediate **M**, which is converted to 2-pyrrolidinone **N** *via* imino enal. Actually, 2-pyrrolidinone **N** could be isolated by the flash column chromatography using silica gel neutralized by trimethylamine. Finally, treatment of **N** with TFA gives the bicyclic lactam **25** as the final product of one-pot reaction cascade.

The aerobic visible light-promoted indole C3 formylation reaction was achieved by using RB as a photocatalyst and *N*,*N*,*N'*,*N'*-tetramethylenediamine (TMEDA) as a one-carbon source (**Figure 11**) [40]. Upon the irradiation of visible light, the reaction of *N*-methylindole **26** with TMEDA in the presence of RB (5 mol%) and KI as an additive under air afforded 3-formyl-*N*-methylindole **27** in 70% yield. This transformation proceeds *via* the addition of *N*-methylindole **26** to iminium ion **O** generated by the oxidation of TMEDA. Next, the C3 thiocyanation reaction of indoles was developed by using ammonium thiocyanate (NH_4SCN) as a thiocyanate radical source [41]. In the presence of RB (1 mol%), the reaction of indole **28** with NH_4SCN gave the adduct **29** in 98% yield. In this reaction, thiocyanate radical is generated by the single-electron transfer between thiocyanate anion and the excited RB (RB*). The thiocyanate radical adds to indole **28**. The subsequent oxidation leads to **29**.

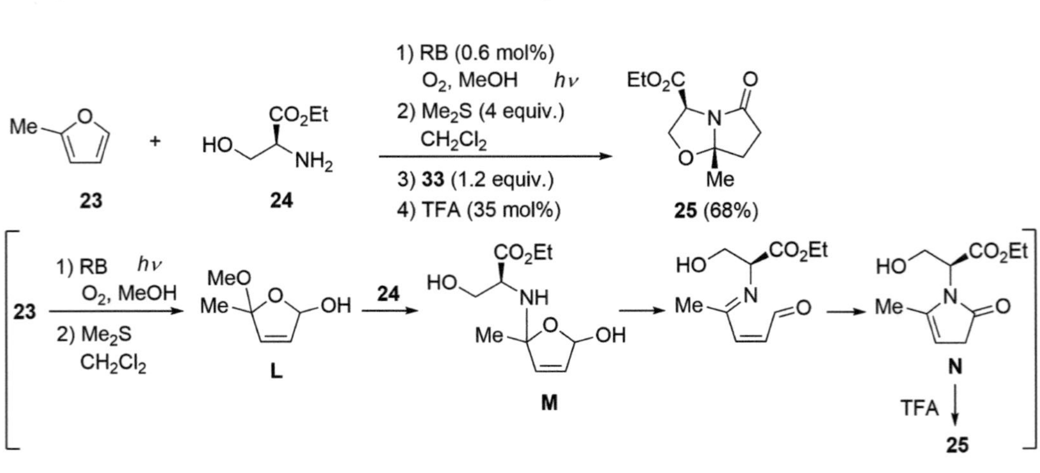

Figure 10. RB-catalyzed synthesis of bicyclic lactam.

http://dx.doi.org/10.5772/intechopen.70507

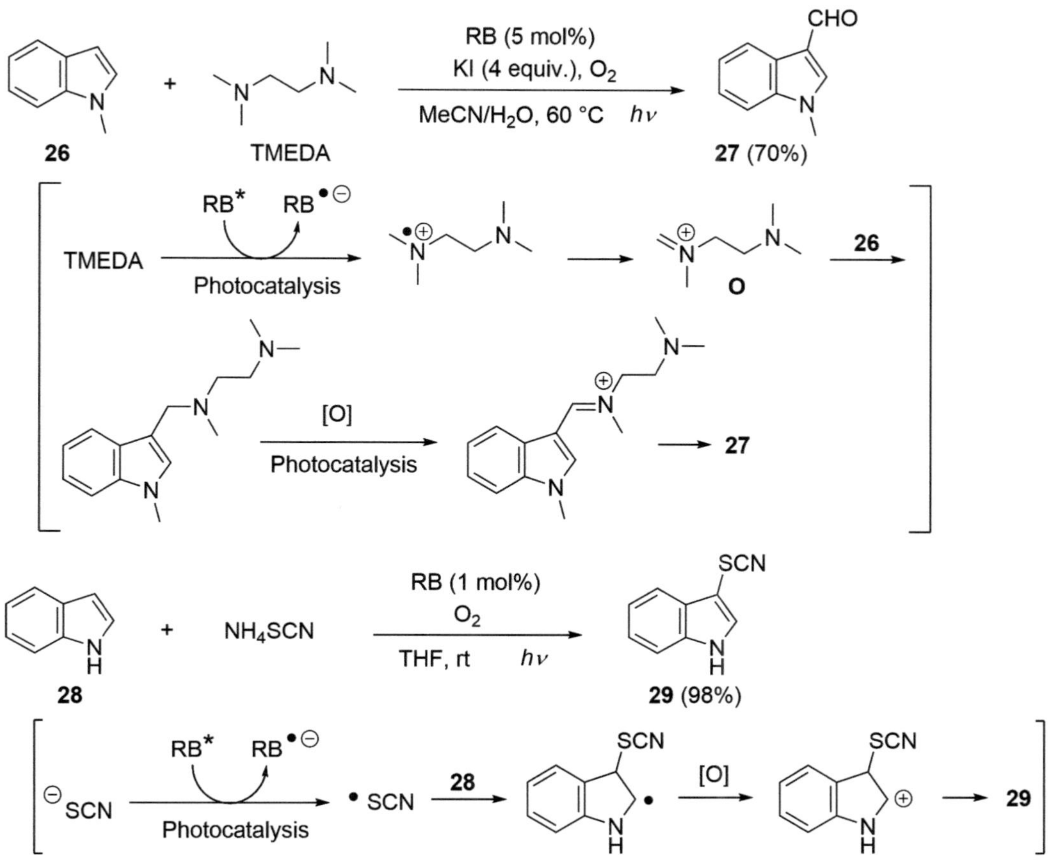

Figure 11. RB-catalyzed functionalization of C3 in indoles.

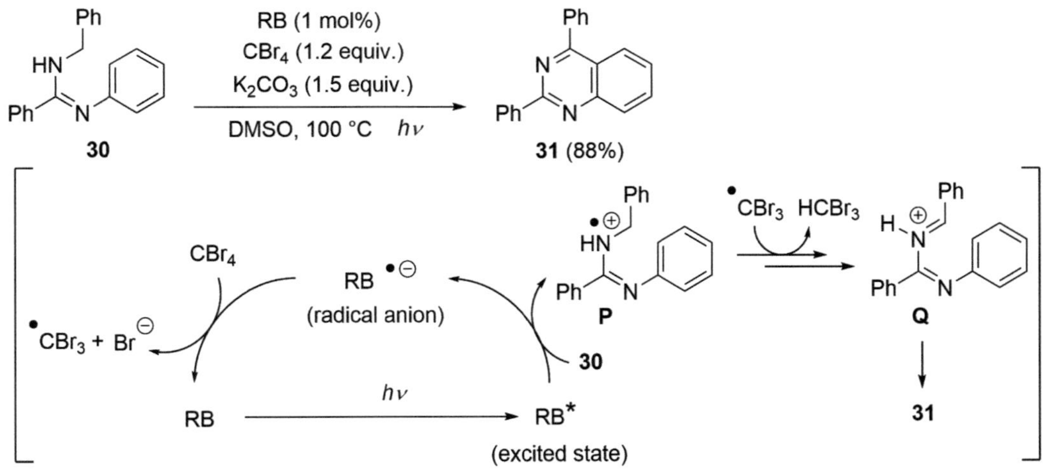

Figure 12. RB-catalyzed oxidative cyclization of benzimidamide.

The new method for the synthesis of quinazolines was developed by using RB as a photocatalyst (**Figure 12**) [42]. In the presence of RB (1 mol%), CBr_4 as an oxidant, and K_2CO_3 as a base, the oxidative carbon-carbon bond-forming cyclization of *N*-benzyl-*N*′-phenyl benzimidamide **30** proceeded smoothly under the visible light irradiation. The desired quinazoline **31** was obtained in 88% yield. In this transformation, CBr_3 radical is generated by the reaction between CBr_4 and the radical anion species of RB. Next, the iminium ion intermediate **Q** is formed from the radical cation intermediate **P** by the association of CBr_3 radical. Finally, the intramolecular Friedel-Craft reaction of iminium ion **Q** leads to quinazoline **31**.

4. Fluorescein and rhodamine B

The utility of fluorescein was demonstrated in the alkoxycarboxylation of aryldiazonium salts using CO gas [45]. In the presence of 0.5 mol% of fluorescein as a photocatalyst, treatment of diazonium tetrafluoroborate **32** with CO (80 atm pressure) in methanol under the irradiation of visible light gave methyl ester **33** in 80% yield (**Figure 13**). Initially, the phenyl radical is generated from diazonium **32** by the single-electron transfer from the excited state of fluorescein (Dye*). Next, benzoyl radical **R** is formed *via* trapping of CO molecule by phenyl radical. The methyl ester **33** is obtained *via* the oxidation of benzoyl radical **R** by the reactive radical cation species of dye followed by trapping of the resulting benzylidyneoxonium **S** with methanol. Furthermore, the dual catalytic system using photocatalyst and gold catalyst was studied by Glorius's group [46].

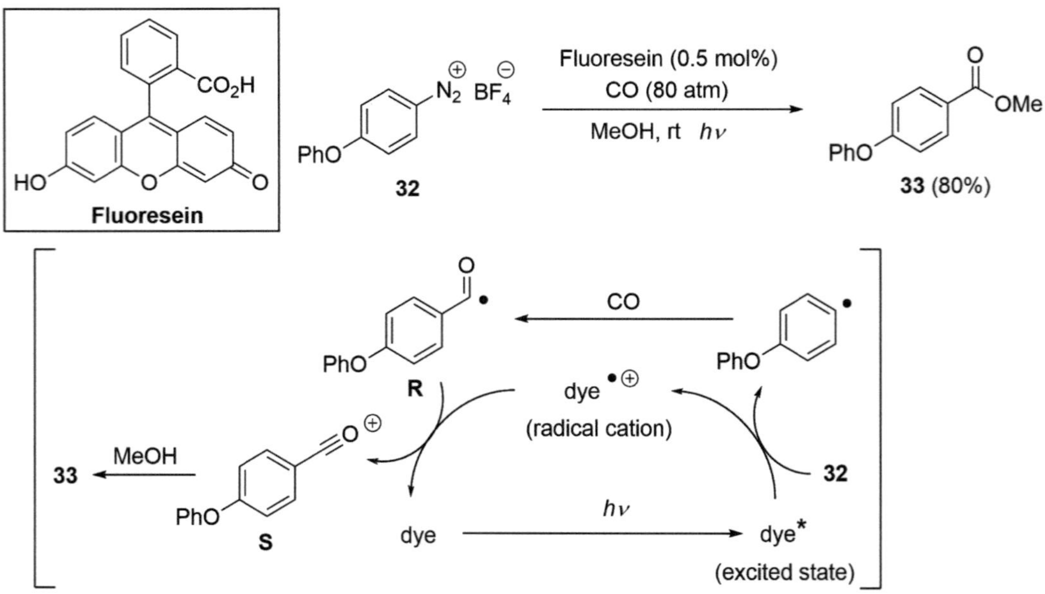

Figure 13. Fluorescein-catalyzed alkoxycarboxylation of aryldiazonium.

The utility of rhodamine B as a water-soluble photocatalyst was demonstrated in the aqueous-medium carbon-carbon bond-forming radical reactions [47]. In the presence of $(i\text{-Pr})_2NEt$ as a reductive quencher, the rhodamine B-catalyzed reaction of alkene **34** with $i\text{-}C_3F_7I$ in H_2O proceeded smoothly to give the product **35** in 90% yield (**Figure 14**). In this transformation, the photo-induced electron transfer from the excited singlet state (S_1) of rhodamine B to $i\text{-}C_3F_7I$ was proposed. This electron transfer process was supported by the fluorescence quenching of rhodamine B with addition of $i\text{-}C_3F_7I$. Additionally, the aqueous-medium radical addition-cyclization-trapping reaction of **36** proceeded effectively even in the absence of $(i\text{-Pr})_2NEt$. In this transformation, the single-electron transfer from iodine ion (I^-) to the radical cation species of rhodamine B in an ion pair would proceed to give I_2, because the oxidation potential of rhodamine B is positive enough to oxidize I^- into I_2.

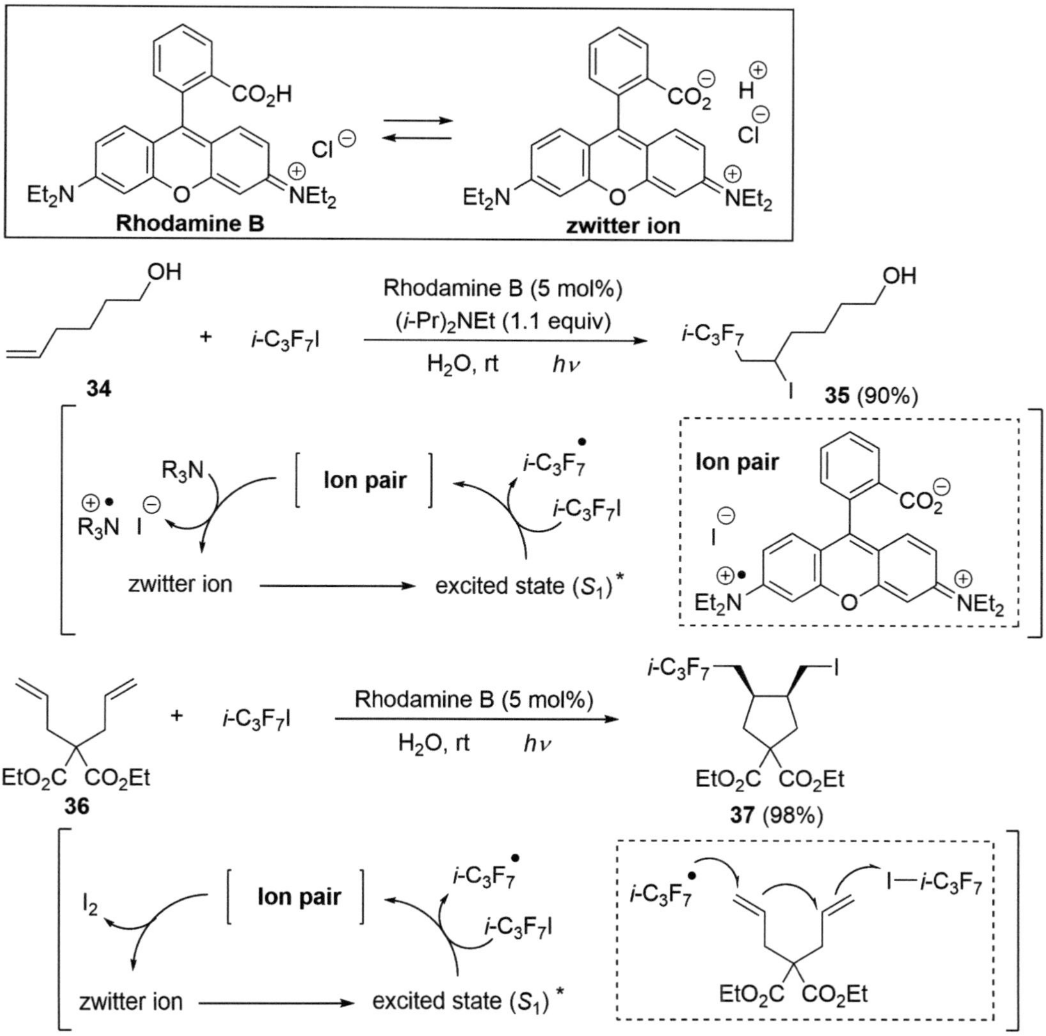

Figure 14. Rhodamine B-catalyzed aqueous-medium radical reactions.

5. Methylene blue and acridine red

Methylene blue (MB) is a member of the thiazine dye family. Scaiano's group used MB as a photocatalyst under the visible light irradiation [48]. The radical trifluoromethylation of electron-rich heterocycles was studied by the use of Togni's reagent **39** as a CF_3 radical source (**Figure 15**). The trifluoromethylation of 3-methylindole **38** proceeded with good yield at low catalyst concentration, when *N,N,N',N'*-tetramethylenediamine (TMEDA) was used as an electron donor. In the presence of 1,8-diazabicyclo[5.4.0]undec-7-ene (DBU) as an electron donor, the reaction of terminal alkene **41** with Togni's reagent **39** also gave the hydrotrifluoromethylation product **42** in 67% yield as a major product, because the fully reduced form of MB, leuco-MB, acts as a hydrogen source [49]. The possible mechanism for the catalytic formation of CF_3 radical is shown. The visible light-excited MB (MB*) is readily quenched by aliphatic amines such as TMEDA or DBU to form the semi-reduced MB as a radical anion and an α-amino radical. CF_3 radical is generated *via* the reduction of Togni's reagent **39** with semi-reduced MB and/or an α-amino radical.

The one-pot transformation of furans into 5-hydroxy-1*H*-pyrrol-2(5*H*)-ones was investigated by using MB as a photocatalyst [50]. In the presence of MB (2 mol%) and oxygen, the reaction of furan **43** with benzylamine gave lactam **44** in 72% yield *via* the reduction of the intermediate by

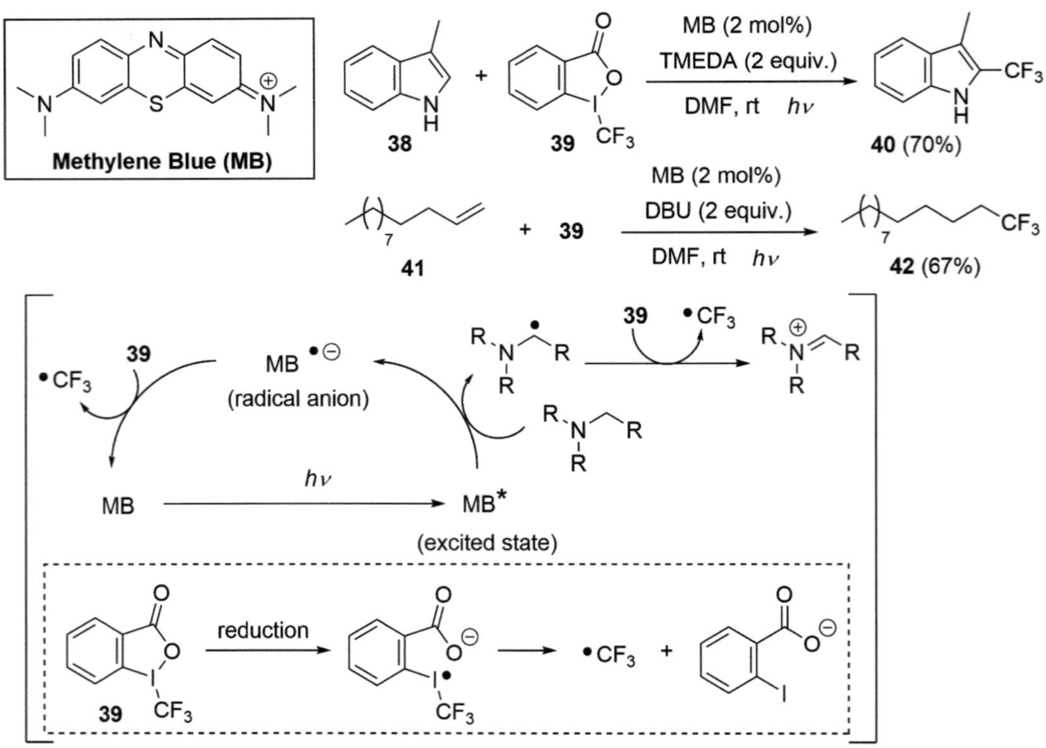

Figure 15. MB-catalyzed trifluoromethylation using Togni's reagent.

Me_2S (**Figure 16**). The use of 2-(3,4-dimethoxyphenyl)ethanamine **45** instead of benzylamine led to the formation of tricycle **46** *via* Pictet-Spengler cyclization process. For Pictet-Spengler cyclization, HCOOH was added as an acid leading to *N*-acyliminium ion, which spontaneously cyclized to form **46**.

New phenothiazine-based organic dye was also developed as a visible light-activated photocatalyst [51].

Acridine red was used as a photocatalyst for the visible light-induced direct thiolation of ethers (**Figure 17**) [52]. The thiolation of tetrahydrofuran (THF) using diphenyl disulfide **47** was carried out in the presence of acridine red (2 mol%) and *tert*-butyl hydroperoxide (TBHP) as an oxidant. The reaction occurred at ambient conditions to give α-arylthioether **48** in 82% yield. This transformation proceeds *via* the generation of radical intermediate from THF, which reacts with diphenyl disulfide **47** to afford 2-(phenylthio)-tetrahydrofuran **48**.

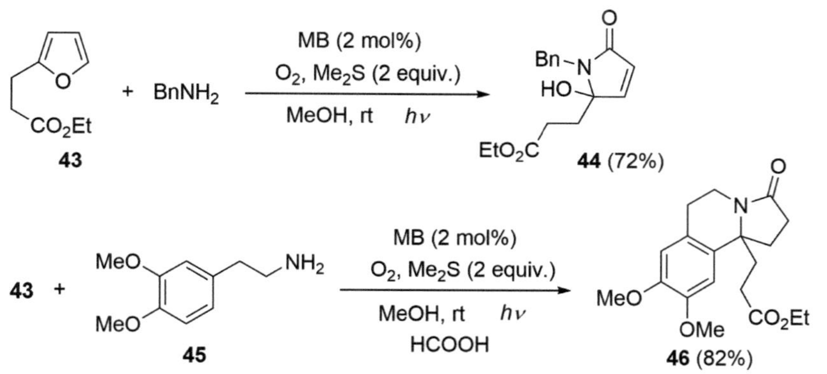

Figure 16. MB-catalyzed reactions of furan with amines.

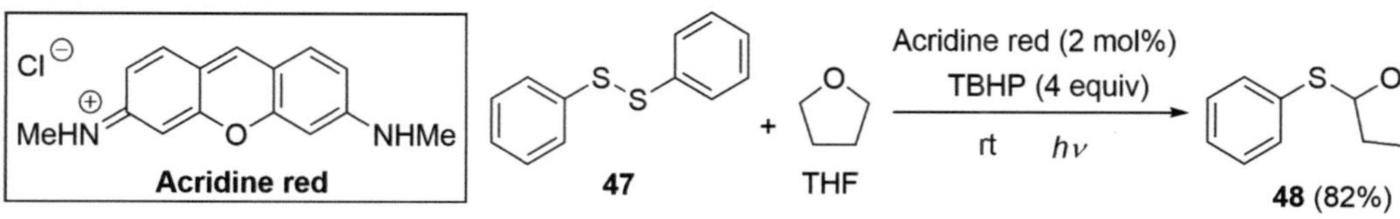

Figure 17. Acridine red-catalyzed thiolation of tetrahydrofuran.

6. Riboflavin tetraacetate

Riboflavin tetraacetate (RFT) is an effective photocatalyst for the visible light-driven organic reactions. The aerobic oxidation of alkyl benzenes to ketones and carboxylic acids was investigated through a dual catalysis using RFT and the tris(2-pyridylmethyl)amine-iron complex $[Fe(TPA)(MeCN)_2](ClO_4)_2$ (TPA=tris(2-pyridylmethyl)amine) [53]. When a mixture of RFT (10 mol%) and $[Fe(TPA)(MeCN)_2](ClO_4)_2$ (2 mol%) was employed, the oxidation of 4-ethylanisole

49 proceeded effectively under the visible light irradiation to give 4-acetylanisole **50** in 80% yield (**Figure 18**). In this oxidation, the iron complex acts as a catalyst for not only oxidation of **49** but also disproportionation of hydrogen peroxide H_2O_2 which is obtained in the RFT-catalyzed oxidation of **49**.

RFT also catalyzed the aerobic oxidation of sulfides to sulfoxides without overoxidation to sulfones [54]. In the presence of RFT (2 mol%), sulfide **51** was transformed chemoselectively to the corresponding sulfoxide **52** in 91% yield (**Figure 19**).

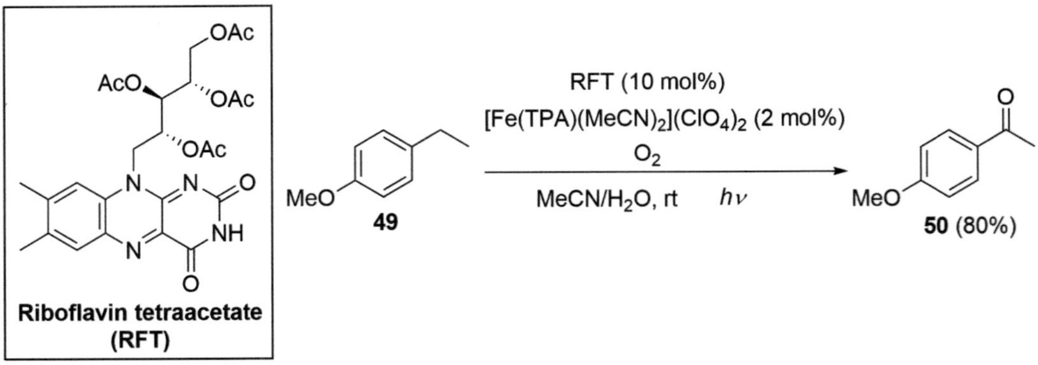

Figure 18. RFT-catalyzed oxidation of 4-ethylanisole.

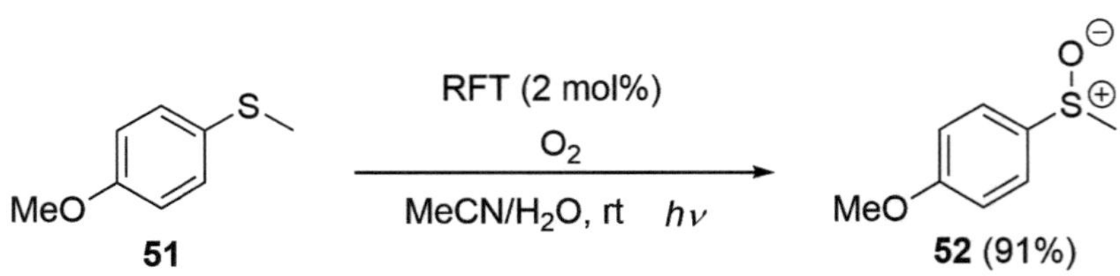

Figure 19. RFT-catalyzed oxidation of sulfide.

7. Concluding remarks

Organic dyes that absorbed visible light induce the synthetically valuable photochemical transformations. The metal-free photocatalysis using organic dyes rapidly progresses in the last few years. In addition to the organic dyes shown in this chapter, Fukuzumi' group has developed 3-cyano-1-methylquinolinium and 9-mesityl-10-methylacridinium ions as photocatalysts [13]. More recently, Nicewicz' group has studied the photocatalysis using acridinium salts [12]. These visible light-induced catalysis disclosed a broader aspect of the utility of organic photocatalysts for synthetic organic chemistry. This chapter will inspire creative new contributions to organic chemists.

Author details

Hideto Miyabe

Address all correspondence to: miyabe@huhs.ac.jp

School of Pharmacy, Hyogo University of Health Sciences, Minatojima, Chuo-ku, Japan

References

[1] Zeitler K. Photoredox catalysis with visible light. Angewandte Chemie International Edition. 2009;**48**:9785-9786

[2] Narayanam JMR, Stephenson CRJ. Visible light photoredox catalysis: Applications in organic synthesis. Chemical Society Reviews. 2011;**40**:102-113

[3] Tucker JW, Stephenson CRJ. Shining light on photoredox catalysis: Theory and synthetic applications. Journal of Organic Chemistry. 2012;**77**:1617-1622

[4] Yoon TP. Visible light photocatalysis: The development of photocatalytic radical ion cycloadditions. ACS Catalysis. 2013;**3**:895-902

[5] Angnes RA, Li Z, Correia CRD, Hammond GB. Recent synthetic Additions to the visible light photoredox catalysis toolbox. Organic and Biomolecular Chemistry. 2015;**13**:9152-9167

[6] Meggers E. Asymmetric catalysis activated by visible light. Chemical Communications. 2015;**51**:3290-3301

[7] Xuan J, Zhang ZG, Xiao WJ. Visible-light-induced decarboxylative functionalization of carboxylic acids and their derivatives. Angewandte Chemie International Edition. 2015;**54**:15632-15641

[8] Amador AG, Yoon TP. A chiral metal photocatalyst architecture for highly enantioselective photoreactions. Angewandte Chemie International Edition. 2016;**55**:2304-2306

[9] Gurry M, Aldabbagh F. A new era for homolytic aromatic substitution: Replacing Bu_3SnH with efficient light-induced chain reactions. Organic and Biomolecular Chemistry. 2016;**14**:3849-3862

[10] Hari DP, König B. Synthetic applications of eosin Y in photoredox catalysis. Chemical Communications. 2014;**50**:6688-6699

[11] Ravelli D, Fagnoni M, Albini A. Photoorganocatalysis. What for? Chemical Society Reviews. 2013;**42**:97-113

[12] Nicewicz DA, Nguyen TM. Recent applications of organic dyes as photoredox catalysts in organic synthesis. ACS Catalysis. 2014;**4**:355-360

[13] Fukuzumi S, Ohkubo K. Organic synthetic transformations using organic dyes as photoredox catalysts. Organic and Biomolecular Chemistry. 2014;**12**:6059-6071

[14] Hari DP, Schroll P, König B. Metal-free, visible-light-mediated direct C-H arylation of heteroarenes with aryl diazonium salts. Journal of the American Chemical Society. 2012;**134**:2958-2961

[15] Meyer AU, Slanina T, Yao CJ, König B. Metal-free perfluoroarylation by visible light photoredox catalysis. ACS Catalysis. 2016;**6**:369-375

[16] Majek M, Jacobi von Wangelin A. Organocatalytic visible light mediated synthesis of aryl sulfides. Chemical Communications. 2013;**49**:5507-5509

[17] Majek M, Jacobi von Wangelin A. Metal-free carbonylations by photoredox catalysis. Angewandte Chemie International Edition. 2015;**54**:2270-2274

[18] Majek M, Filace F, Jacobi von Wangelin A. Visible light driven hydro-/deuterodefunctionalization of anilines. Chemistry A European Journal. 2015;**21**:4518-4522

[19] Meng QY, Zhong JJ, Liu Q, Gao XW, Zhang HH, Lei T, Li ZJ, Feng K, Chen B, Tung CH, Wu LZ. A cascade cross-coupling hydrogen evolution reaction by visible light catalysis. Journal of the American Chemical Society. 2013;**135**:19052-19055

[20] Zhong JJ, Wu CJ, Meng QY, Gao XW, Lei T, Tung CH, Wu LZ. A cascade cross-coupling and *in situ* hydrogenation reaction by visible light catalysis. Advanced Synthesis & Catalysis. 2014;**356**:2846-2852

[21] Meyer AU, Jäger S, Hari DP, König B. Visible light-mediated metal-free synthesis of vinyl sulfones from aryl sulfinates. Advanced Synthesis & Catalysis. 2015;**357**:2050-2054

[22] Yang W, Yang S, Li P, Wang L. Visible-light initiated oxidative cyclization of phenyl propiolates with sulfinic acids to coumarin derivatives under metal-free conditions. Chemical Communications. 2015;**51**:7520-7523

[23] Gu L, Jin C, Liu J, Ding H, Fan B. Transition-metal-free, visible-light induced cyclization of arylsulfonyl chlorides with 2-isocyanobiphenyls to produce phenanthridines. Chemical Communications. 2014;**50**:4643-4645

[24] Davies J, Booth SG, Essafi S, Dryfe RAW, Leonori D. Visible-light-mediated generation of nitrogen-centered radicals: Metal-free hydroimination and iminohydroxylation cyclization reactions. Angewandte Chemie International Edition. 2015;**54**:14017-14021

[25] Yang Z, Li H, Zhang L, Zhang MT, Cheng JP, Luo S. Organic photocatalytic cyclization of polyenes: A visible-light-mediated radical cascade approach. Chemistry A European Journal. 2015;**21**:14723-14727

[26] Jadhav SD, Bakshi D, Singh A. Visible light mediated organocatalytic activation of ethyl bromofluoroacetate: Coupling with indoles and anilines. Journal of Organic Chemistry. 2015;**80**:10187-10196

[27] Zhang J, Wang L, Liu Q, Yang Z, Huang Y. Synthesis of α,β-unsaturated carbonyl compounds *via* a visible-light-promoted organocatalytic aerobic oxidation. Chemical Communications. 2013;**19**:11662-11664

[28] Zhou C, Li P, Zhu X, Wang L. Merging photoredox with palladium catalysis: Decarboxylative *ortho*-acylation of acetanilides with α-oxocarboxylic acids under mild reaction conditions. Organic Letters. 2015;**17**:6198-6201

[29] Li J, Zhang J, Tan H, Wang DZ. Visible-light-promoted vinylation of tetrahydrofuran with alkynes through direct C-H bond functionalization. Organic Letters. 2015;**17**:2522-2525

[30] Mitra S, Ghosh M, Mishra S, Hajra A. Metal-free thiocyanation of imidazoheterocycles through visible light photoredox catalysis. Journal of Organic Chemistry. 2015;**80**: 8275-8281

[31] Luo K, Chen YZ, Yang WC, Zhu J, Wu L. Cross-coupling hydrogen evolution by visible light photocatalysis toward C(sp^2)-P formation: Metal-free C-H functionalization of thiazole derivatives with diarylphosphine oxides. Organic Letters. 2016;**18**:452-455

[32] Liu H, Feng W, Kee CW, Zhao Y, Leow D, Pan Y, Tan CH. Organic dye photocatalyzed α-oxyamination through irradiation with visible light. Green Chemistry. 2010;**12**:953-956

[33] Pan Y, Kee CW, Chen L, Tan CH. Dehydrogenative coupling reactions catalysed by rose Bengal using visible light irradiation. Green Chemistry. 2011;**13**:2682-2685

[34] Pan Y, Wang S, Kee CW, Dubuisson E, Yang Y, Loh KP, Tan CH. Graphene oxide and rose Bengal: Oxidative C-H functionalization of tertiary amines using visible light. Green Chemistry. 2011;**13**:3341-3344

[35] Liu H, Feng W, Kee CW, Zhao Y, Leow D, Pan Y, Tan CH. Organic dye photocatalyzed α-oxyamination through irradiation with visible light. Green Chemistry. 2012;**14**:953-956

[36] Teo YC, Pan Y, Tan CH. Organic dye-photocatalyzed acylnitroso ene reaction. Chem CatChem. 2013;**5**:235-240

[37] Kalaitzakis D, Montagnon T, Alexopoulou I, Vassilikogiannakis G. A versatile synthesis of Meyers' bicyclic lactams from furans: Singlet-oxygen-initiated reaction cascade. Angewandte Chemie International Edition. 2012;**51**:8868-8871

[38] Kalaitzakis D, Antonatou E, Vassilikogiannakis G. One-pot synthesis of 1-azaspiro frameworks initiated by photooxidation of simple furans. Chemical Communications. 2014;**50**:400-402

[39] Triantafyllakis M, Tofi M, Montagnon T, Kouridaki A, Vassilikogiannakis G. Singlet oxygen-mediated synthesis of bis-spiroketals found in azaspiracids. Organic Letters. 2014;**16**:3150-3153

[40] Li X, Gu X, Li Y, Li P. Aerobic transition-metal-free visible-light photoredox indole C-3 formylation reaction. ACS Catalysis. 2014;**4**:1897-1900

[41] Fan W, Yang Q, Xu F, Li P. A visible-light-promoted aerobic metal-free C-3 thiocyanation of indoles. Journal of Organic Chemistry. 2014;**79**:10588-10592

[42] Shen Z, Yang P, Tang Y. Transition metal-free visible light-driven photoredox oxidative annulation of arylamidines. Journal of Organic Chemistry. 2016;**81**:309-317

[43] Fidaly K, Ceballos C, Falguières A, Veitia MSI, Guy A, Ferroud C. Visible light photoredox organocatalysis: A fully transition metal-free direct asymmetric α-alkylation of aldehydes. Green Chemistry. 2012;**14**:1293-1297

[44] Barata-Vallejo S, Yerien DE, Postigo A. Benign perfluoroalkylation of aniline derivatives through photoredox organocatalysis under visible-light irradiation. European Journal of Organic Chemistry. 2015;**2015**:7869-7875

[45] Guo W, Lu LQ, Wang Y, Wang YN, Chen JR, Xiao WJ. Metal-free, room-temperature, radical alkoxycarbonylation of aryldiazonium salts through visible-light photoredox catalysis. Angewandte Chemie International Edition. 2015;**54**:2265-2269

[46] Hopkinson MN, Sahoo B, Glorius F. Dual photoredox and gold catalysis: Intermolecular multicomponent oxyarylation of alkenes. Advanced Synthesis & Catalysis. 2014;**356**: 2794-2800

[47] Yoshioka E, Kohtani S, Jichu T, Fukazawa T, Nagai T, Kawashima A, Takemoto Y, Miyabe H. Aqueous-medium carbon–carbon bond-forming radical reactions catalyzed by excited rhodamine B as a metal-free organic dye under visible light irradiation. Journal of Organic Chemistry. 2016;**81**:7217-7229

[48] Pitre SP, McTiernan CD, Ismaili H, Scaiano JC. Metal-free photocatalytic radical trifluoromethylation utilizing methylene blue and visible light irradiation. ACS Catalysis. 2014;**4**:2530-2535

[49] Yamamoto S, Fujiyama Y, Shiozaki M, Sueishi Y, Nishimura N. Hydride transfer reactions of leuco methylene blue and leuco thionine with some *p*-benzoquinones. Journal of Physical Organic Chemistry. 1995;**8**:805-809

[50] Kalaitzakis D, Kouridaki A, Noutsias D, Montagnon T, Vassilikogiannakis G. Methylene blue as a photosensitizer and redox agent: Synthesis of 5-hydroxy-1*H*-pyrrol-2(5*H*)-ones from furans. Angewandte Chemie International Edition. 2015;**54**:6283-6287

[51] Park JH, Ko KC, Kim E, Park N, Ko JH, Ryu DH, Ahn TK, Lee JY, Son SU. Photocatalysis by phenothiazine dyes: Visible-light-driven oxidative coupling of primary amines at ambient temperature. Organic Letters. 2012;**14**:5502-5505

[52] Zhu X, Xie X, Li P, Guo J, Wang L. Visible-light-induced direct thiolation at α-C(sp^3)-H of ethers with disulfides using acridine red as photocatalyst. Organic Letters. 2016;**18**:1546-1549

[53] Mihldorf B, Wolf R. C-H photooxygenation of alkyl benzenes catalyzed by riboflavin tetraacetate and a non-heme iron catalyst. Angewandte Chemie International Edition. 2016;**55**:427-430

[54] Neveselý T, Svobodová E, Chudoba J, Sikorski M, Cibulka R. Efficient metal-free aerobic photooxidation of sulfides to sulfoxides mediated by a vitamin B_2 derivative and visible light. Advanced Synthesis & Catalysis. 2016;**358**:1654-1663

Chapter 2

Rice Husk for Photocatalytic Composite Material Fabrication

Diana Rakhmawaty Eddy, Atiek Rostika Noviyanti, Solihudin Solihudin, Safri Ishmayana and Roekmi-ati Tjokronegoro

Additional information is available at the end of the chapter

http://dx.doi.org/10.5772/intechopen.72704

Abstract

As a semiconductor, zinc oxide (ZnO) has better UV absorbing properties compared to other semiconductor materials, and therefore, it has better dye degrading abilities. However, ZnO tends to agglomerate, which lead to poor degradation compared to the other semiconductors. In this study, to overcome the agglomeration of ZnO, silica (SiO_2) was combined with ZnO. The composite was tested for its photocatalytic activity. The ZnO/SiO_2 photocatalyst was fabricated on a glass plate. In order to investigate the addition of SiO_2 on ZnO, X-ray diffraction (XRD) and scanning electron microscope-energy dispersive X-ray spectroscopy (SEM-EDS) was used. The result of the XRD analysis demonstrates similar peak results with ZnO XRD data from ICSD 157132 with a hexagonal structure. The results indicate that the ZnO structure did not change after the addition of SiO_2, while SEM-EDS results showed that SiO_2 was supported on ZnO with 8% composition. The optimal composition was found to be ZnO/SiO_2 95/5, as indicated by high degradation activity, which can degrade up to 89% methylene blue.

Keywords: rice husk, silica, photocatalysis, ZnO, methylene blue

1. Introduction

The textile industry is developing at a rapid pace, and this has a positive impact on garments development. However, it also increases the negative impact through their industrial waste, especially textile dye. One of the means to degrade dye is by the use of a semiconductor material that has photocatalytic activity [1]. Utilization of semiconductors in photocatalysis is an interesting topic, attributable to its ability to degrade compounds with ultraviolet light facilitation [2].

TiO_2 is usually used as a photocatalyst because it is stable compared to the other photocatalytic agents. However, TiO_2 absorbs less UV light compared to ZnO. Therefore, ZnO can degrade more dye. However, in reality, ZnO degrades less dye compared to TiO_2, since ZnO tends to agglomerate [3]. To overcome this problem, SiO_2 can be added to ZnO. Pure ZnO degrades 40% dye in 60 minutes. However, when SiO_2 was added, the degradation increased and showed better results compared to pure ZnO. The addition of SiO_2 to ZnO achieved optimum photocatalytic activity at ZnO/SiO_2 90/10 weight ratio with 99% dye degradation efficiency [4].

Rice husk is composed of about 20% paddy grain [5]. It is composed of mainly cellulose (~32%), silica (~22%) and lignin (~16%) [6]. After milling, rice husk becomes a major waste product, and is not utilized optimally. In 2015, Indonesia produced 75 million tonnes of rice (Central Agency on Statistics, bps.go.id), and therefore, rice husk becomes abundant and a cheap source for silica.

The efficiency of degradation and decolourization is affected by the stability of ZnO layer and ZnO layer morphology. An experiment was conducted by fabricating ZnO/SiO_2 composite with 95/5, 90/10, and 85/15 weight ratio. The results indicated good results when the composite was used to degrade methylene blue with synthetic SiO_2 made from rice husk as the supporting material. Therefore, the present study aims to fabricate ZnO/SiO_2 with less agglomeration [4].

2. Experimental section

2.1. Isolation of SiO_2

Rice husk was carbonated at 400°C for 6 hours, followed by increasing the heat to 700°C in an argon atmosphere for 4 hours. The carbon was then ground and seized using a 100 mesh sieve. The carbon was suspended in with a mole ratio 1/3/150 = silica:potassium carbonate: water and refluxed for 150 minutes. The mixture was filtered and the filtrate was allowed to cool. The SiO_2 was precipitated and collected after filtering the solution.

2.2. Fabrication of ZnO/SiO_2 nanocomposite

Three weight ratio of ZnO/SiO_2, namely, 100/0, 95/5, 90/10 and 85/15, were prepared to give 3 g of total mass. The solid mixture was then suspended in 100 mL of distilled water. The suspension was stirred using a magnetic stirrer (500 rpm) for 2 hours, followed by sonication for 90 minutes (Elma ultrasonic LC 30H). The ZnO/SiO_2 suspension was dropped onto a glass slide (1 × 3 cm) with a pipette until the entire glass surface was covered. The slide was dried at 40°C for 12 hours, followed by calcinations at 450°C for 1 hour. The slide was washed using distilled water, and the layer was characterized using X-ray diffraction (XRD) and scanning electron microscope-energy dispersive X-ray spectroscopy (SEM-EDS).

2.3. Determination of maximum wavelength and standard curve

The absorbance of 1 ppm methylene blue solution was recorded at 500–700 nm in wavelength. The wavelength at which the highest absorbance was detected was used to measure the concentration of methylene blue. The standard curve was made by measuring the absorbance of methylene blue solutions with concentration values of 0.2, 0.4, 0.6 and 0.8 ppm. The absorbance was plotted against concentration, and used as a standard curve.

2.4. Photocatalytic assay

For photocatalytic assay, a glass side with ZnO/SiO_2 layer was inserted into 50 mL of 1 ppm methylene blue in a test tube. It was followed by irradiation using a mercury lamp for 4 hours. Every 2 hours, 2 mL of sample was collected, and its methylene blue content was determined using a visible spectrophotometer at 660 nm. The assay was performed for all the fabricated slides.

3. Results and discussions

3.1. ZnO/SiO_2 photocatalyst

The ZnO/SiO_2 photocatalyst was fabricated to give final mass of 3 g with 95/5, 90/10, 85/15 weight ratio, as described in previous studies [4, 7]. The fabrication was conducted in distilled water, while assay was performed in two different solvents, i.e., distilled water and methanol. According to the assay results, both solvents yielded the same ZnO attachment level on the glass slide. Therefore, both water and methanol are effective as solvents, as previously proposed [7]. In order for ZnO and SiO_2 to completely disperse in water, the suspension was stirred with a magnetic stirrer at 500 rpm for 2 hours, as recommended by prior work [4].

Sonication with ~30 kHz for 90 minutes was conducted to homogenize the ZnO/SiO_2 so that no agglomeration occurs. Prior to the coating, the glass slide was cleaned using acetone to remove impurities that can interfere with the attachment of ZnO/SiO_2. The glass slide that was coated with ZnO/SiO_2 was dried in a 40°C oven for 12 hours to remove excess water so that the ZnO/SiO_2 attaches strongly onto the glass slide (**Figure 1**).

To increase the attachment of ZnO/SiO_2 onto the glass slide, it was heated to 450°C. The surface where ZnO/SiO_2 is attached should be a flat surface, so that when the surface is washed it is easier for the agglomerated ZnO/SiO_2 to be washed [4].

3.2. Characterization results

SEM-EDS and XRD was performed to investigate the ZnO coating onto glass slides. Complete analysis was performed for ZnO/SiO_2 95/5 weight ratio. The result of SEM is illustrated in **Figure 2**. It can be observed on the 500× magnification that the ZnO/SiO_2 layer attached uniformly with low porosity. A porous structure started to appear at 1000× magnification. The particles attachment affects the efficiency of methylene blue degradation, since it determines hydroxyl radical generation by ZnO to degrade the dye.

Figure 3 shows the ZnO layer without SiO_2 addition. It appears that the particles do not attach uniformly, and have a higher porosity compared to the ZnO/SiO_2 layer, as shown in **Figure 2**. It appears that the addition of SiO_2 to ZnO is significantly effective to facilitate a uniform spread

Figure 1. ZnO/SiO_2 and ZnO.

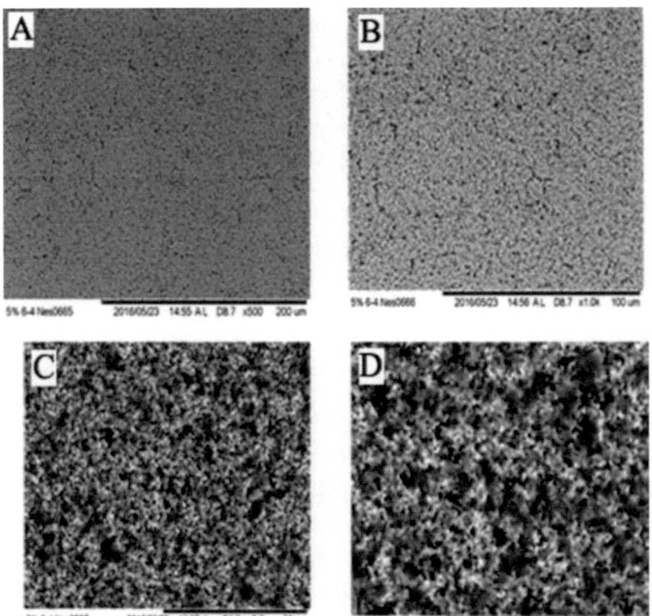

Figure 2. Micrograph of ZnO/SiO_2 at 95/5 weight ratio at (A) 500-, (B) 1000-, (C) 2500-, and (D) 5000-fold magnification.

of ZnO when attached onto the glass slide surface, which can then reduce the porosity of the ZnO/SiO_2 layer. A lower porosity leads to a higher ZnO attached to the glass slide, which acts as a hydroxyl radical generator. Based on the comparison results of ZnO to ZnO/SiO_2 with 95/5 weight ratio, it is clear that the composite can produce more hydroxyl radical, which leads to a more effective dye degradation.

The EDS results are presented in **Figure 4**. EDS is used to investigate the composition of ZnO and SiO_2 in ZnO/SiO_2 mixture. With 95/5 weight ratio, Zn and Si composition was 61.64 and 8.38% weight respectively. However, according to the calculation results, the composition of Zn and Si should be 76.23 and 2.30% weight respectively. A higher Si composition and a lower Zn composition compared to the theoretical value is due to the ZnO being carried

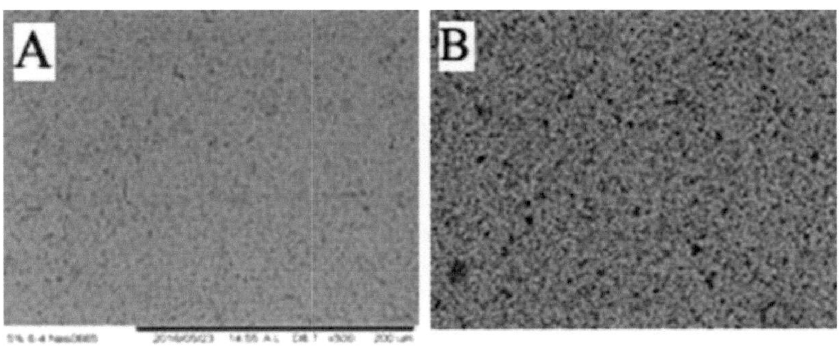

Figure 3. Micrograph of (A) ZnO/SiO_2 95/5 weight ratio and (B) ZnO with 500× magnification.

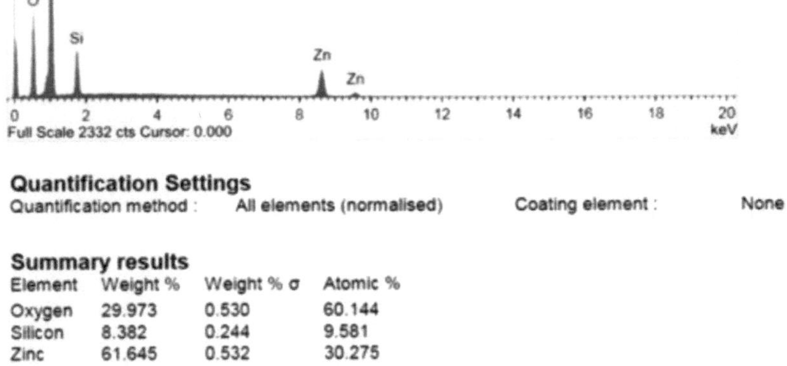

Figure 4. EDS results of ZnO/SiO_2 95/5 weight ratio.

away during the washing process, especially the agglomerated ZnO. The agglomeration may occur when sonication is extensively long. When more ZnO is washed away, a more porous structure occurs on the layer (**Figure 3**). A high porosity on the layer reduces the ability of ZnO to form hydroxyl radical, which then reduces the rate of degradation. From the EDS results (**Figure 3**), it was proven that 8.38% of SiO_2 is supported on the ZnO particles. It was also found that around 1.22% weight was not detected by the EDS method, which indicates the presence of impurities in the layer.

To investigate the effect of SiO_2 on the ZnO structure, XRD was used. The result of XRD is presented in **Figure 5**. It was found that peaks at 32.33°, 34.67° and 36.85° of ZnO/SiO_2 was shifted compared to standard ZnO. This shift occurred as a result of the SiO_2 attachment on ZnO. The intensity of the peaks indicates that there is no difference between peaks of ZnO/SiO_2 compared to standard ZnO (ICSD 157132 standard), which has a hexagonal crystal structure. This result is in agreement with previously published results, which describe that layers of ZnO have a hexagonal structure.

The results indicate that the addition of SiO_2 does not change the ZnO crystal structure. However, the addition of SiO_2 can increase the distribution of ZnO on the glass slide, thus resulting in a more homogen layer, which leads to no agglomeration. The XRD of ZnO/SiO_2 shows that the SiO_2 peak appears with very low intensity at 2θ = 20–25° (**Figure 5**). The low intensity peak appears because the SiO_2 amount in ZnO/SiO_2 was low, i.e. 5%. The peak at 2θ = 23.5° with low intensity belongs to SiO_2 from rice husk with an amorphous property.

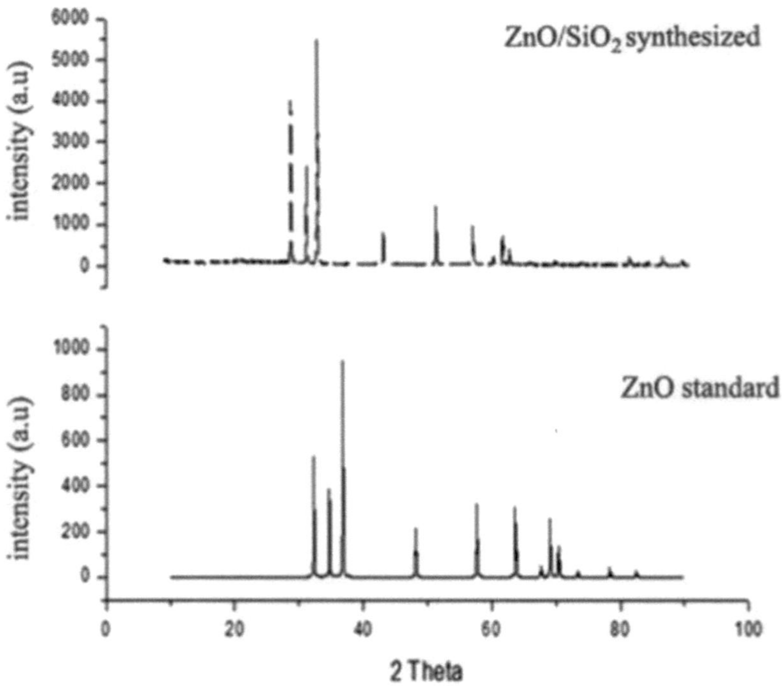

Figure 5. XRD of ZnO/SiO_2 and ZnO standard (ICSD 157132).

This result confirmed that SiO_2 used to fabricate ZnO/SiO_2 is SiO_2 from rice husk with an amorphous property (**Figure 6**).

3.3. Maximum wavelength and standard curve of methylene blue

The wavelength at which maximum absorbance of methylene blue was detected is 660 nm. The maximum wavelength has the highest sensitivity, and therefore the measurement of methylene blue concentration was conducted at this wavelength to the minimize error rate. The correlation of the standard curve was 0.9909, which shows that the method has good correlation.

3.4. Photocatalysis assay

In the photocatalysis assay, samples were collected four times every hour, and the concentration of methylene blue was determined by measuring the absorbance of the samples and converted to a concentration using a standard curve. The percentage of the efficiency of methylene blue degradation was obtained by comparing the concentration of methylene blue before and after radiation. The efficiency of degradation is presented in **Figure 7**.

As shown in **Figure 7**, after 1 hour of radiation, ZnO/SiO_2 with 95/5 weight ratio was more effective compared to the other samples, as it can degrade up to 52.5% of methylene blue. After 3 hours of irradiation, ZnO/SiO_2 with 90/10 weight ratio degrades the dye better compared to the other samples. After 4 hours of irradiation, ZnO/SiO_2 photocatalysts with weight ratio of 95/5. 90/10 and 85/15 showed 89.95, 82.57 and 37.40% degradation percentage respectively. The photocatalysts with 95/5 and 90/10 weight ratio is more effective in degrading methylene blue dye compared to 85/15 weight ratio, as indicated by percentage degradation above 80%. When SiO_2 was used by more than 10%, the ability of the photocatalysts became lower. This is likely caused by SiO_2 covering ZnO that should be act as photocatalysts, which leads to a lower hydroxyl radical formation. ZnO/SiO_2 photocatalysts with 85/15 weight ratio

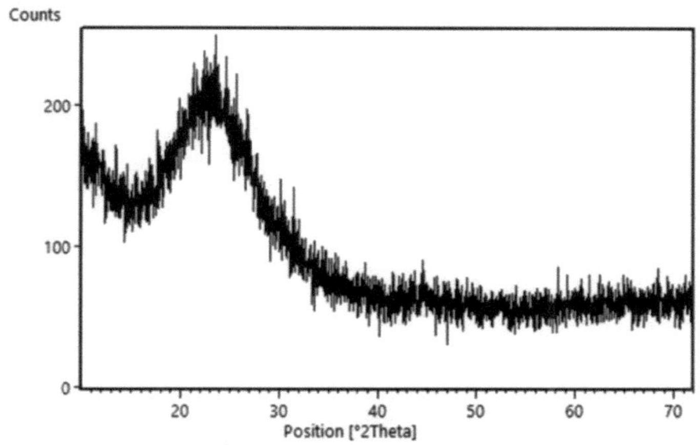

Figure 6. XRD of SiO_2.

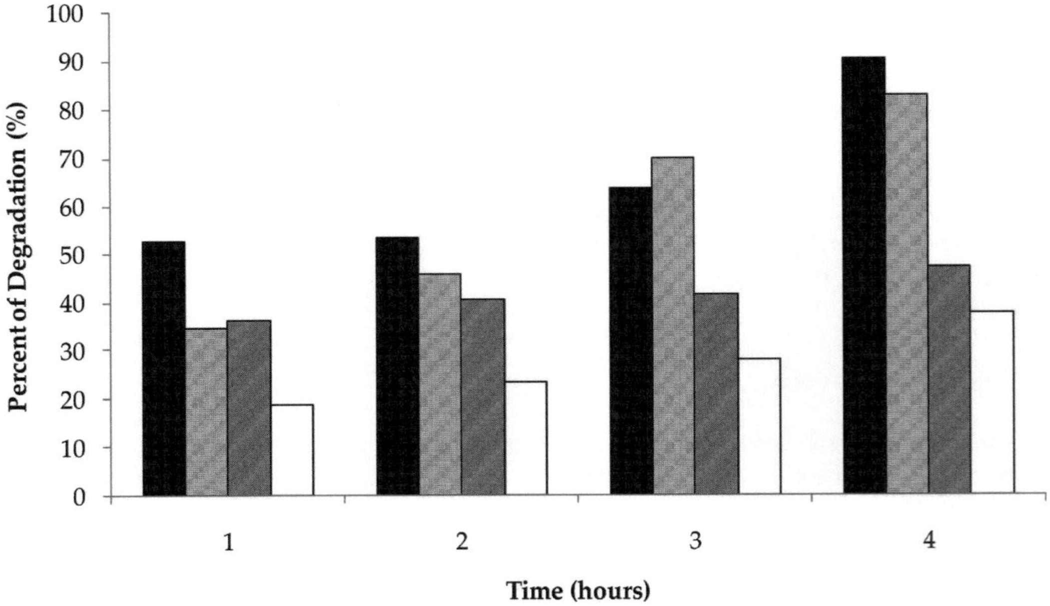

Figure 7. Efficiency of methylene blue degradation using ZnO/SiO_2 photocatalysis with various ZnO/SiO_2 weight ratio. Irradiation time was 4 hours.

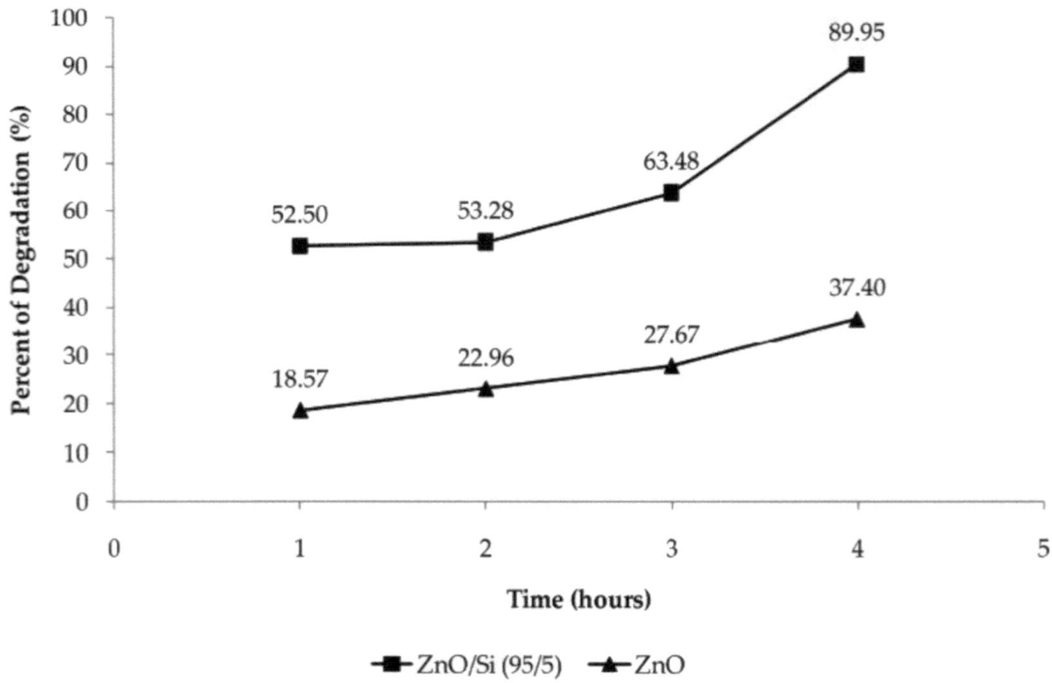

Figure 8. Comparison of methylene blue degradation efficiency using ZnO/SiO_2 compared to ZnO photocatalyst.

less efficient because SiO_2 is not a semiconductor material, so it cannot produce free electrons and hydroxyl radicals. According to Behnajady et al. [8], less hydroxyl radicals will lead to a lower dye reduction efficiency. **Figure 7** confirms this, where photocatalysts with 95/5 weight ratio showed the highest dye degradation.

Figure 8 shows the comparison between ZnO/SiO_2 and ZnO photocatalytic activity. It is clear that the addition of SiO_2 increases the effectiveness of the phototocatalyst, as indicated by higher methylene blue degradation of up to 89.95%, while ZnO without SiO_2 can only degrade 37.40%. According to Soltani et al. (2015), this is caused because the ZnO attachment without SiO_2 was not uniform on the glass slide surface that tends to form a porous structure. The addition of SiO_2 to ZnO assisted the attachment to form a uniform layer, and this leads to lower porosity.

4. Conclusion

The addition of silica from rice husk to ZnO photocatalysts improves the spread of the particles uniformly on the glass slide. The addition of the silica does not alter the crystal structure of the ZnO. The optimal composition was found to be ZnO/SiO_2 95/5, as indicated by a high degradation activity, which can degrade up to 89% methylene blue.

Author details

Diana Rakhmawaty Eddy*, Atiek Rostika Noviyanti, Solihudin Solihudin, Safri Ishmayana and Roekmi-ati Tjokronegoro

*Address all correspondence to: diana.rahmawati@unpad.ac.id

Department of Chemistry, Faculty of Mathematics and Natural Sciences, Universitas Padjadjaran, Jatinangor, Sumedang, Indonesia

References

[1] Selvam K, Swaminathan K, Chae K-S. Decolourization of azo dyes and a dye industry effluent by a white rot fungus *Thelephora* sp. Bioresource Technology. 2003;**88**:115-119

[2] Dijkstra MF, Buwalda H, de Jong AWF, Michorius A, Winkelman JGM, Beenackers AACM. Experimental comparison of three reactor designs for photocatalytic water purification. Chemical Engineering Science. 2001;**56**:547-555

[3] Behnajady MA, Moghaddam SG, Modirshahla N, Shokri M. Investigation of the effect of heat attachment method parameters at photocatalytic activity of immobilized ZnO nanoparticles on glass plate. Desalination. 2009;**249**:1371-1376

[4] Soltani RD, Shams Khoramabadi G, Godini H, Noorimotlagh Z. The application of ZnO/SiO_2 nanocomposite for the photocatalytic degradation of a textile dye in aqueous solutions in comparison with pure ZnO nanoparticles. Desalination and Water Treatment. 2015;**56**:2551-2558

[5] Madrid R, Nogueira CA, Margarindo F. Production and characterization of amorphous silica from rice husk waste. In: The 4th International Conference on Engineering for Waste and Biomass Valorisation; 10-13 September 2012; Porto, Portugal; 2012

[6] Prasad R, Pandey M. Rice husk ash as a renewable source for the production of value added silica gel and its application: An overview. Bulletin of Chemical Reaction Engineering & Catalysis. 2012;**7**(1):1-25

[7] Behnajady MA, Modirshahla N. Photooxidative degradation of Malachite Green (MG) by UV/H_2O_2: Influence of operational parameters and kinetic modeling. Dyes and Pigments. 2006;**70**:54-59

[8] Behnajady MA, Modirshahla N, Daneshvar N, Rabbani M. Photocatalytic degradation of C.I. Acid Red 27 by immobilized ZnO on glass plates in continuous-mode. Journal of Hazardous Materials. 2007;**140**:257-263

Chapter 3

Silica Gel–Supported P-, Ge-, and Sb-Porphyrins for Visible Light Inactivation of Bacteria

Masahide Yasuda and Jin Matsumoto

Additional information is available at the end of the chapter

http://dx.doi.org/10.5772/intechopen.70004

Abstract

This chapter describes the photocatalysis action of (dihydroxo)tetraphenyl¬porphyrinato complexes of high valent P (V), Ge (IV), and Sb (V) (P(tpp), Ge(tpp), and Sb(tpp)). These chromophores were fixed onto silica gel (SiO2) through Coulombic forces and hydrogen bonding between axial hydroxo ligands and silanol groups to produce M(tpp)/SiO2 (M = P, Ge, and Sb) composites. M(tpp)/SiO2 were applied to the photo-inactivation of *Escherichia coli* and *Legionella pneumophila*. Moreover, M(tpp)/SiO2 was subjected to practical experiments for the photoinactivation of *L. pneumophila* naturally occurring in a cooling tower and a public fountain. It is noteworthy that 80 g of Sb(tpp)/SiO2 catalyst, containing 40 mg of Sb(tpp) maintained a concentration of Legionella species below 100 CFU/100 mL for 120 days in 13 m3 of water in a fountain under sunlight exposure. The photoinactivation proceeded through the liberation of M(tpp) from SiO2, adsorption of M(tpp) inside bacteria, and generation of reactive oxygen species, such as singlet oxygen, under visible light irradiation, thus resulting in bacteria apoptosis. Based on these results, we developed water-soluble porphyrins by modification of P and Sb porphyrin axial ligands to alkyloxo, alkylethylenedioxy, and alkylpyridinium groups. These water-soluble porphyrins were applied to the photodynamic inactivation of E. coli and *Saccharomyces cerevisiae*.

Keywords: high valent metal, P(V)-porphyrin, Ge(IV)-porphyrin, Sb(V)-porphyrin, *Escherichia coli, Legionella pneumophila, Saccharomyces cerevisiae*

1. Introduction

Photocatalysis has received much attention as an environmentally friendly process for degrading organic compounds and bacteria in contaminated water. It is well known that UV-light irradiation of TiO_2 generates a hydroxyl radical, which works as a strong oxidizing reagent for various microorganisms in aqueous solutions [1]. In 1985, Matsunaga et al. reported that Pt-loaded TiO_2, under irradiation of >380 nm light, was capable of photoinactivation of various microorganisms, such as *Lactobacillus acidophilus, Saccharomyces cerevisiae, Escherichia coli,* and *Chlorella vulgaris* [2]. However, the UV light used for activation of TiO_2 is harmful to humans and only low amounts of the correct wavelengths are found in sunlight. Therefore, a photocatalyst that operates under irradiation from visible light emitted from the sun or a fluorescent lamp has been earnestly desired. Many researchers have focused much attention on porphyrins and metalloporphyrins, which can absorb visible light with high-absorption coefficients. Among the various metalloporphyrin chromophores, we selected axially dihydroxo-substituted tetraphenylporphyrin complexes (M(tpp)) of high valent metals, such as P(V), Ge(IV), and Sb(V) (**Figure 1**). To support M(tpp) on a carrier, we used silica gel (SiO_2), which has a high transparency to visible light and a strong binding force to porphyrins. Thus, M(tpp)/SiO_2 composites were prepared by adsorption of M(tpp) on SiO_2 and the composites operated as photocatalysts under visible light irradiation. Moreover, the porphyrins were subjected to water solubilization for medical field applications.

In this chapter, we will cover M(tpp)/SiO_2 and water-soluble porphyrins that were applied to the photosensitized inactivation of *E. coli* and *S. cerevisiae* in a reaction vessel and *Legionella pneumophila* in naturally occurring environments such as cooling tower and public fountains.

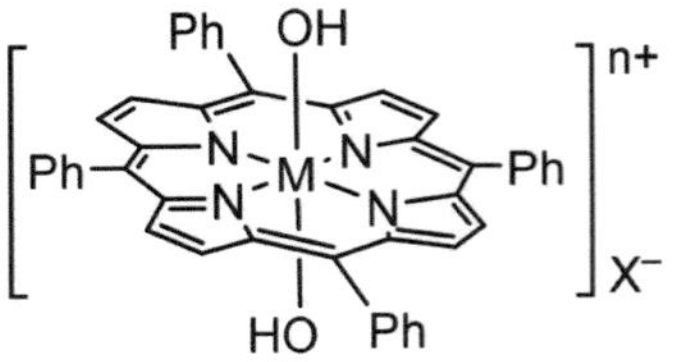

P(tpp); M = P, *n* = 0, X = Cl
Ge(tpp); M = Ge, *n* = 0, X = none
Sb(tpp); M = Sb, *n* = 1, X = Br

Figure 1. P(tpp), Ge(tpp), and Sb(tpp).

2. Preparation of the photocatalysts

2.1. Porphyrin chromophores

The porphyrin chromophores studied were (dihydroxo) tetraphenylporphyrinatophosphorus chloride (P(tpp)), (dihydroxo)tetraphenylporphyrinatogermanium (Ge(tpp)), and (dihydroxo)tetraphenylporphyrinatoantimony bromide (Sb(tpp)). Axial hydroxo ligands were bonded to the metals through stable covalent bonds. P(tpp) and Sb(tpp) are cationic complexes, whereas Ge(tpp) is a neutral complex. **Table 1** lists the physicochemical properties of the M(tpp)s, such as oxidation potentials ($E_{1/2}^{ox}$ vs. Ag/$AgNO_3$), reduction potentials ($E_{1/2}^{red}$), absorption maxima (λ_{max}), and molar absorption coefficients (ε). Since M(tpp) has absorptions at the Soret band around 420 nm and Q-bands at 550 nm with high ε, they can capture even weak light emitted from a fluorescent lamp and the sun. The fluorescence maxima (λ_F), quantum yields (Φ_F), and lifetimes (τ_F) were obtained from fluorescence spectra. Since Φ_F values were low, it was suggested that the efficiency of triplet state formation was relatively

M(tpp)	$E_{1/2}^{ox}$/V[a]	$E_{1/2}^{red}$/V[b]	λ_{max}/nm[c] (ε /10^4 M^{-1} cm^{-1})[d]		Fluorescence		Ref.
			Soret band	Q-band	λ_{max}/(E^{0-0}/eV)[e]	Φ_F (τ_F/ns)[f]	
P(tpp)	1.20[g]	−0.93[g]	424 (31.2)	554 (1.82)	607 (2.04)	0.0416 (5.4)	[3]
Ge(tpp)	0.95	−0.83	420 (77.6)	554 (2.19)	596 (2.08)	0.1500 (4.7)	[4, 5]
Sb(tpp)	1.17	−0.74	419 (41.6)	552 (3.09)	596 (2.08)	0.0518 (1.6)	[6, 7]

[a]Half peak of oxidation potential *vs* $Ag/AgNO_3$.
[b]Half peak of reduction potential *vs* $Ag/AgNO_3$.
[c]Absorption maxima of M(tpp).
[d]Molar absorption coefficients (ε) of M(tpp) in MeCN. The ε of P(tpp) was measured in MeOH.
[e]Fluorescence maxima. The values in parenthesis are excitation energy in eV.
[f]Fluorescence quantum yield (Φ_F) under excitation at the Q-band and fluorescence lifetimes (τ_F).
[g]Unpublished results.

Table 1. Properties of M(tpp).

high. High efficiency of triplet state formation is advantageous for energy transfer to $^3O_2$. The $E_{1/2}^{red}$ values of M(tpp)s were −0.74 to −0.93 V vs. $Ag/AgNO_3$, which were relatively positive compared with the divalent metal tetraphenylporphyrin complexes of Zn(II) (−1.31 V), Ni(II) (−1.18 V), and Pb(II) (−1.10 V) [8]. This fact shows that M(tpp)s (M = P, Ge, Sb) have powerful oxidation abilities under light irradiation. Moreover, the toxicity of M(tpp) is low. The lethal dose 50 (LD_{50}) of Sb(tpp) [9] and P(tpp) [10] was more than 2000 mg/kg. Therefore, M(tpp) can be safely used in living environmental fields.

2.2. Preparation of porphyrins/silica gel composites

To perform the photoreaction in an aqueous solution, less water-soluble M(tpp) was fixed onto porous SiO_2 to form M(tpp)/SiO_2 [11]. Silica gel powder (300 mesh, 40 μmϕ, BW300 Fuji Silysia, Japan) and silica gel beads (0.85–1.70 mmϕ, 306 m^2 g^{-1}, CARIACT Q–10, Fuji Silysia) were used. M(tpp) was fixed on SiO_2 through Coulombic forces and hydrogen bonding between axial hydroxo ligands and silanol groups. The general procedure for M(tpp)/SiO_2 preparation is described for the case of Sb(tpp)/SiO_2 as follows. SiO_2 (70 g) was added to a MeOH–toluene solution (1:4 v/v, 500 mL) containing Sb(tpp) (60 mg) and the mixture was allowed to stand for 18 h. MeOH was evaporated from the solution at 40°C under reduced pressure (**Figure 2**). The treated silica gel was isolated by filtration and then dried under reduced pressure at 40°C to give Sb(tpp)/SiO_2. The content of Sb(tpp) was 0.087 wt%.

2.3. Photocatalytic oxidation of organic compounds using M(tpp)/SiO_2 (M = Ge, Sb)

Since M(tpp)/SiO_2 (M = Ge, Sb) have high oxidation abilities, M(tpp)/SiO_2 was applied to the oxidation of cycloalkenes [12] and acetone [4] and the dechlorination of 4-chlorophenol [13]. The photocatalytic reactions were performed using the setup depicted in **Figure 3**, where the reactant was supplied by a continuous flow system. In a spiral-type apparatus (**Figure 3A**), the reactant solution is fed continuously from a holder to a spiral glass tube packed with the

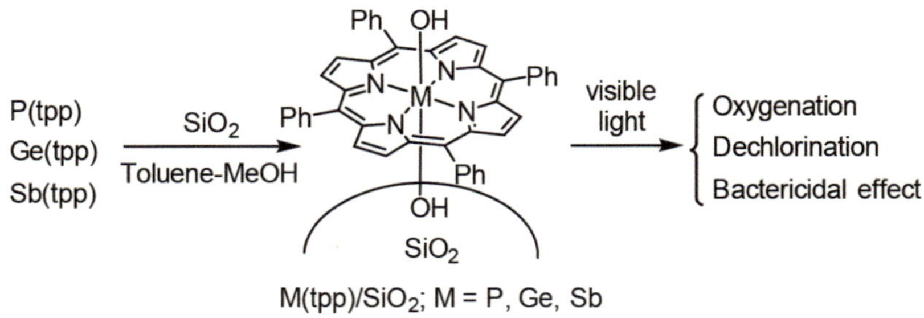

Figure 2. Silica gel-supported M(tpp) catalyst (M(tpp)/SiO_2).

photocatalyst. Sample irradiation was performed using a fluorescent lamp (22 W). The oval mirror-type apparatus (**Figure 3B**) consisted of a fluorescent lamp (18 W), an oval mirror, and a reactor (20 mmϕ × 500 mm; 150 mL) packed with the photocatalyst. The fluorescent lamp was set on one focus of the oval mirror, and the reactor was set at another focus. The visible light emitted from the fluorescent lamp was concentrated onto the reactor. The reactant solution was fed continuously from reservoirs into the reactors.

The photocatalytic oxidations of cycloalkenes using oxygen were performed in a spiral type reactor (**Figure 3A**). Irradiation was directed onto a spiral glass tube (4 mmΦ × 2.5 m) containing photocatalyst (6.0 g) to which CH_2Cl_2 solutions (150 mL) of cycloalkenes (0.1 M) were fed continuously from the reservoir [12]. To reduce silanol group effects, Sb(tpp)/SiO_2 was modified by capping with $(Me_3Si)_2NH$ to give Sb(tpp)/SiO_2^{TMS}. The photocatalytic oxidation of cyclohexene on Sb(tpp)/SiO_2^{TMS} in CH_2Cl_2 produced *cis*-1,2-epoxycyclohexane (28%), 1,2-cyclohexanediol (27%), and 2-cyclohexen-1-ol (38%) along with a small amount of 2-cyclohexene-1-one (7%) at 99.8% conversion after irradiation for 56 h (**Figure 4**). Similarly, the photocatalytic oxidation was applied to cyclooctene and 1-methyl-cyclohexene, which mainly resulted in diol formation.

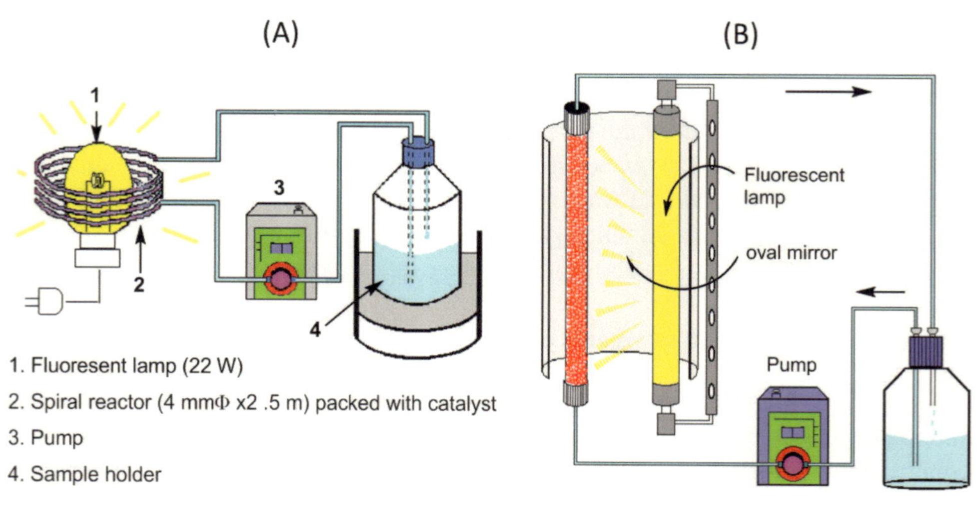

Figure 3. Setups for photocatalytic reactions: spiral type (A) and oval mirror type (B).

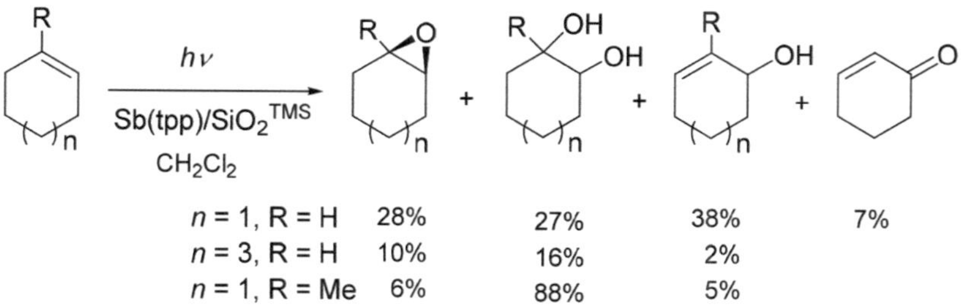

Figure 4. Photo-oxidation of cycloalkenes using $Sb(tpp)/SiO_2^{TMS}$.

Photo-oxidation of MeOH to HCHO was performed by $Ge(tpp)/SiO_2$ at room temperature using a spiral type reactor (**Figure 3A**). Irradiation with a fluorescent light was performed on a spiral glass tube (4 mmϕ × 2.5 m) packed with $Ge(tpp)/SiO_2$ (12 g). The reactant was fed continuously into the spiral glass tube from the reservoir that contained an aerated aqueous solution of MeOH (50 mM in 200 mL) [4]. HCHO (30.6 μM) was formed after 180 h with a turnover number (TON) of 3.0 (**Figure 5**). The isotope effect for the photo-oxidation of methanol was found to be 2.1 from the ratios of slopes in the time-conversion plots for CH_3OH and CD_3OH. Therefore, it was suggested that the oxidation occurs through hydrogen abstraction. The generation of a Ge-O• species was generated by excitation of Ge(tpp). Here, Ge(tpp) acts as an O-radical generator. The photo-oxidation by $Ge(tpp)/SiO_2$ was applied to toluene and ethylbenzene to produce alcohols and aldehydes/ketones. In the cases of cumene, methylcyclopentane, and methylcyclohexane, the corresponding alcohols were produced.

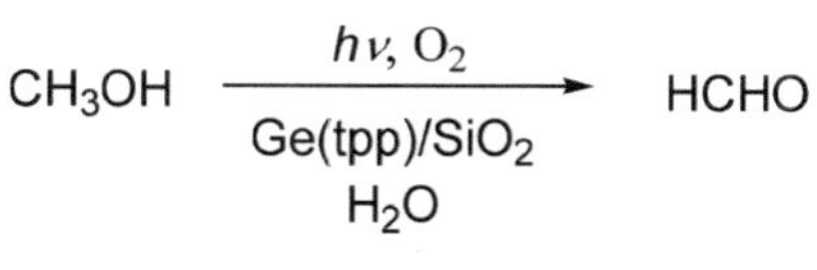

Figure 5. Photo-oxidation of methanol using $Ge(tpp)/SiO_2$.

Dechlorination of 4-chlorophenol (4-CP) using $Sb(tpp)/SiO_2$ was performed using the oval mirror-type apparatus (**Figure 3B**) [13]. Here, $Fe(NO_3)_3$ was used instead of O_2 as the electron acceptor for the dechlorination of 4-CP, since the oxidation potential of 4-CP was relatively high. $Sb(tpp)/SiO_2$ (0.087 wt% Sb(tpp)) was loaded into the oval mirror reactor. Before irradiation, the aqueous solution of 4-CP (initial concentration was 493 μM) was fed for 3 h under dark conditions, and the initial concentration of 4-CP decreased to 400 mM, probably due to the adsorption of 4-CP on the SiO_2. Upon irradiation with the fluorescent lamp for 72 h, the concentration of 4-CP decreased from 400 to 6 μM along with the formation of Cl^- (233 μM) and 1,4-benzoquinone (205 μM). Fe^{2+} (811 μM) was produced as a consequence of the reduction of Fe^{3+} (**Figure 6**). Electron transfer from the excited triplet state of $Sb(tpp)/SiO_2$ to Fe^{3+} was responsible for the photodechlorination initiation, since Rehm-Weller equation calculated that the free energy change (ΔG) for the electron transfer from the excited triplet state of Sb(tpp) (E^{0-0} = 1.63 eV) to Fe^{3+} ion ($E_{1/2}^{red}$ = −0.31 V vs $Ag/AgNO_3$) would be −0.15 V: $\Delta G = E_{1/2}^{ox} - E_{1/2}^{red} - E^{0-0}$, where $E_{1/2}^{ox}$ (Sb(tpp)) = 1.17 V vs $Ag/AgNO_3$. The resulting reactive

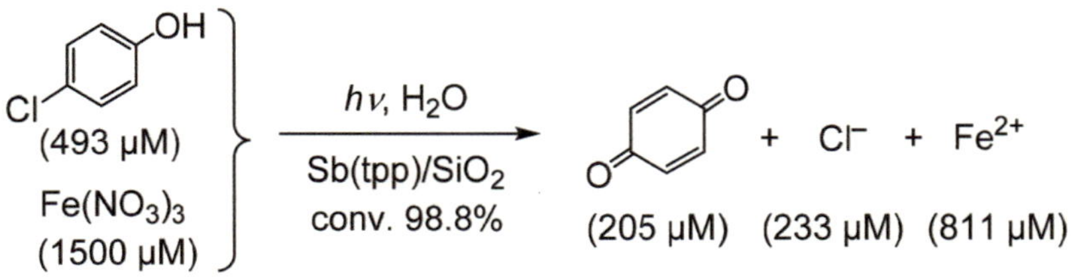

Figure 6. Dechlorination of 4-chlorophenol (4-CP) using Sb(tpp)/SiO_2 under irradiation from a fluorescent lamp.

dicationic Sb(tpp)/SiO_2 undergoes the hole transfer to 4-CP adsorbed on SiO_2 and solved in the aqueous solution. The resulting 4-CP cation radical allows for the nucleophilic addition of H_2O. The substitution of Cl with OH via the hydroxy adduct followed by the oxidation by Fe^{3+} gives 1,4-benzoquinone.

3. Photoinactivation of *E. coli* by M(tpp)/SiO_2 (M = P, Sb)

Our first experiment for photoinactivation using Sb(tpp)/SiO_2 was reported in 2003 against *E. coli* [10]. Photoinactivation of *E. coli* was enacted as follows: *E. coli* k-12 (IFO3335) was cultured aerobically at 30°C for 8 h in a basal medium (pH 6.5) consisting of 1% bactotriptone, 0.5% yeast extract, and 1% NaCl. After centrifugation of the cultured broth at 8500 g for 10 min, the harvested cells were washed with physiological saline and then resuspended in the saline. Phosphate buffer (9.0 mL, 100 mM, pH 7.0), the cell suspension of *E. coli* (1.0 mL, ca. 10^4 cells mL^{-1}), and Sb(tpp)/SiO_2 (10 mg) were introduced into an L-type glass tube (length 180 mm, diameter 15 mm) that was set on a reciprocal shaker in the apparatus shown in

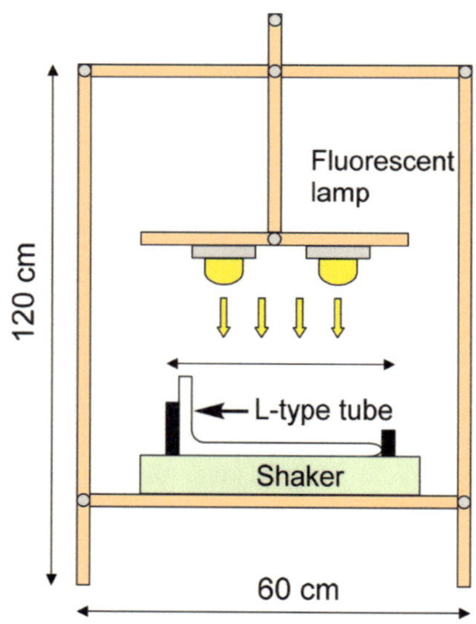

Figure 7. Apparatus used for photoinactivation of bacteria experiments.

http://dx.doi.org/10.5772/intechopen.70004

Figure 7. The glass tube was shaken at 160 rpm and irradiated for 2 h using a fluorescent lamp (Panasonic FL-15ECW, Japan, λ = 400–723 nm, maximum intensity at 545 nm, 10.5 W cm^{-2}) set above a reciprocal shaker. Aliquots (0.1 mL) were taken from the reaction mixture at 20 min intervals and plated on agar plates in triplicate. The colonies that appeared after incubation for 24 h at 30°C were counted. The amount of living cells (*B* in cell mL^{-1}) was defined as the average number of the colonies of *E. coli* in the triplicate plates.

Figure 8A shows the time courses of survival ratio (100 B/B_0) of *E. coli*, where B_0 is the initial amount of *E. coli*. Upon irradiation for 60 min, the concentration of *E. coli* decreased from the initial concentration (6.4×10^3 cells mL^{-1}) to about 40 cells mL^{-1}. Its survival ratio was 0.6%. Thus, Sb(tpp)/SiO_2 demonstrated bactericidal activity under visible light irradiation. Control experiments, irradiation in the absence of the photocatalyst and photocatalyst without irradiation, were also performed. Each control run maintained the B_0 of *E. coli*, showing that the Sb(tpp)/SiO_2 has bactericidal activity only with light activation. Incident light was absorbed exclusively by Sb(tpp)/SiO_2. However, the excited singlet state of Sb(tpp)/SiO_2 was too short-lived to contact directly with the bacteria. Therefore, it was suggested that the excited triplet state of Sb(tpp)/SiO_2 (triplet energy = 1.63 eV) underwent energy transfer to molecular oxygen (triplet energy = 0.98 eV) to generate singlet oxygen ($^1O_2$). Thus, we found that Sb(tpp)/SiO_2 could generate $^1O_2$ in aqueous solutions to sterilize *E. coli* cells using visible light irradiation.

Similar photoinactivation studies of *E. coli* were performed using P(tpp)/SiO_2 (10 mg) in phosphate buffer (10 mL) containing *E. coli* (10^4 cells) [10]. The survival plots are shown in **Figure 8B**. Under a nitrogen atmosphere, photoinactivation using P(tpp)/SiO_2 did not occur. We postulated that photoinactivation obeyed a Michaelis-Menten type mechanism, which involves an interaction between bacteria and P(tpp)/SiO_2 in the ground state. Plots of B_0–B

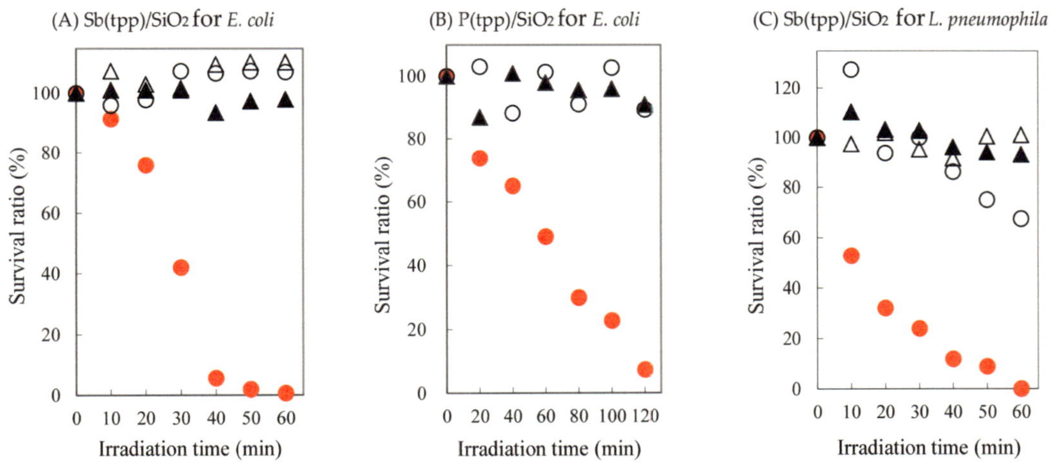

Figure 8. Photoinactivation of *E. coli* by Sb(tpp)/SiO_2 (A) and P(tpp)/SiO_2 (B) and *L. pneumophila* by Sb(tpp)/SiO_2 (C) under visible light irradiation in the presence of M(tpp)/SiO_2 (●), dark conditions in the presence of M(tpp)/SiO_2 (▲), visible light irradiation in the absence of M(tpp)/SiO_2 (○), and dark in the absence of M(tpp)/SiO_2 (△). Reaction conditions: initial concentration of bacteria = 1.0×10^4 cell mL^{-1} for *E. coli* and 6.4×10^3 cells mL^{-1} for *L. pneumophila*, M(tpp)/SiO_2 = 10 mg in phosphate buffer solution (10 mL) for *E. coli* and water (10 mL) for *L. pneumophila* under irradiation of fluorescent lamps (light intensity = 21 W cm^{-2}).

values against irradiation time gave a linear correlation. From the slopes of the plots, the rate for photoinactivation was determined to be proportional to the amounts of P(tpp)/SiO_2 and light intensity. Thus, adsorption of *E. coli* on P(tpp)/SiO_2 was important for efficient photoinactivation using P(tpp)/SiO_2 under visible light irradiation.

4. Photoinactivation of *Legionella* species in naturally occurring environments

L. pneumophila has become one of the most aggressive pathogens since the first outbreak of Legionnaires' disease in Philadelphia in 1976 [14]. The natural habitats for *L. pneumophila* include a wide range of aquatic bodies, such as lakes, streams, and artificially constructed aquatic reservoirs (hot springs, fountains, and cooling towers). Large outbreaks of Legionnaires' disease have been reported in Portugal, the Netherlands, and Spain [15–17]. On the other hand, photocatalytic treatments have received much attention as an environmentally friendly process to inactivate bacteria in contaminated water. In 2002-2003, our group carried out the photoinactivation of *Legionella* species occurring in bacteria's natural habitats, such as in cooling towers and public fountains using a use of Sb(tpp)/SiO_2 [18].

4.1. Photoinactivation of *L. pneumophila*

Initially, the photoinactivation of *L. pneumophila* was examined with Sb(tpp)/SiO_2 using L-type glass tubes in the apparatus shown in **Figure 7** in a similar manner to the method described for *E. coli* [10]. A phosphate buffer (0.1 M, pH 7.0, 10 mL) containing a cell suspension of *L. pneumophila* (6.4×10^5 CFU/100 mL), and Sb(tpp)/SiO_2 (10 mg) was introduced into the L-type glass tube (length 18 cm, diameter 1.5 cm) and irradiated with a fluorescent lamp [18]. Aliquots (0.1 mL) of the reaction mixture were directly plated on a selective medium for *Legionella* species, i.e., WYOα agar medium (Eiken Chemicals Co., Ltd, Japan) consisting of glycine (3 g), vancomycin (5 mg), polymixin B (10^6 IU), and amphotericin B (80 mg). The colonies of *L. pneumophila* appeared after incubation for 7 days at 36°C. Wet, smooth, and bluish-white colonies were counted on triplicate plates. The cell concentration of *L. pneumophila* was represented in colony formation units in 100 mL of the aqueous solution (CFU/100 mL).

Figure 8C shows the time courses for *Legionella* species survival. Upon irradiation for 60 min in the presence of the Sb(tpp)/SiO_2, the concentration of *Legionella* species apparently decreased from the initial concentration (6.4×10^5 CFU) to 4×10^3 CFU. Its survival ratio was 0.6%. In the control experiments, irradiation in the absence of Sb(tpp)/SiO_2, in the presence of Sb(tpp)/SiO_2 without irradiation, and in the absence of Sb(tpp)/SiO_2 and irradiation, each runs maintained the initial concentration of *Legionella* species. Thus, Sb(tpp)/SiO_2 was confirmed to have the photocatalytic activity that could sterilize *Legionella* species.

4.2. Practical experiments in a cooling tower

The bactericidal experiment was performed in a cooling tower (**Figure 9A**) that was located in a building in Miyazaki city [18]. A cylindrical apparatus (200 mmϕ × 500 mm, **Figure 9B**)

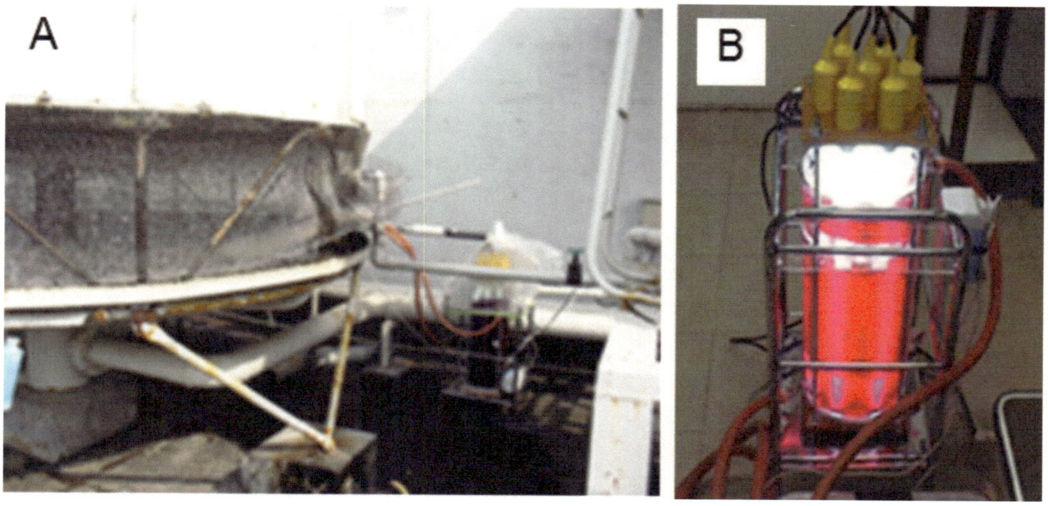

Figure 9. Practical experiment in a cooling tower: the cylindrical apparatus (B) setup in the cooling tower (A).

that consisted of seven fluorescent lamps (18 W, 43 mmϕ × 50 cm) and the Sb(tpp)/SiO_2 catalyst (4.0 kg, 0.05 wt% Sb(tpp)) was used. Water in the holder (800 L) of the cooling tower was pumped into a cylindrical vessel at a rate of 28 L min^{-1}, and then, the treated water was returned to the holder. The average retention time was calculated to be 26 s. In the cylindrical vessel, the Sb(tpp)/SiO_2 catalyst was irradiated by visible light emitted from the fluorescent lamps at ambient temperature. Sampling of the water was carried out at outlet at 3–7 day intervals. At the same time, the atmospheric temperatures were recorded as the average value of the highest temperature of Miyazaki city during the sampling day and 2 days prior. The sample water (1.0 L) was filtrated through a membrane filter (0.45 μm, HA, Millipore) under reduced pressure into the vessel (100 mL) containing the microbes adhering to the membrane filter, an aqueous solution (5 mL) was added, and the vessel was shaken vigorously. Saturated aqueous KCl (5 mL, pH 2.2) containing 0.2 M HCl was added, and the vessel was shaken vigorously. After standing for exactly 20 min at room temperature, the prepared solution was ready for plating as described in Section 4.1. The amounts of *Legionella* species were determined by the colony counting method.

Under the conditions without any bactericidal treatments, *Legionella* species occurred in a range from 20 to 139 CFU in the holder of the cooling tower, as shown in **Figure 10**. After the bactericidal apparatus was active, the concentrations of *Legionella* species decreased to levels below the detection limit. This was maintained until the irradiation treatment ceased. Seven days after the irradiation ceased, detectable amounts of *Legionella* species reappeared. Thus, the bactericidal effects of Sb(tpp)/SiO_2 were practically confirmed in this cooling tower experiment.

4.3. Practical experiments in water fountain

Practical experiments were performed in a public fountain of Miyazaki city (**Figure 11A**) that was filled with 13 m^3 of water [18]. Photoinactivation of the fountain was examined using a

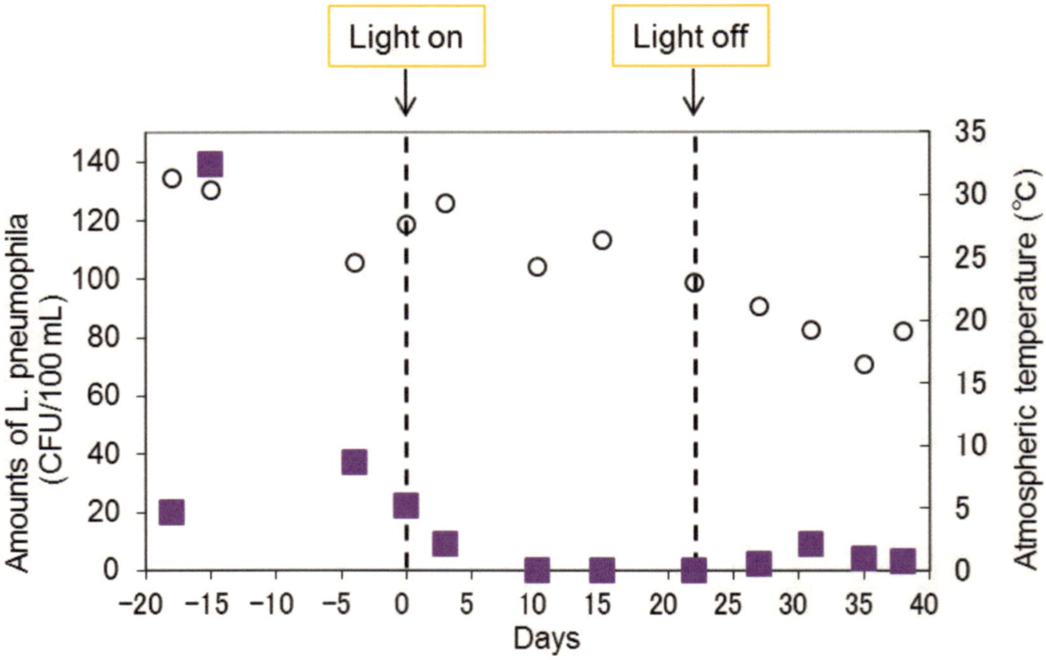

Figure 10. Time-course plots of the amounts of *Legionella* species (■) in the cooling tower along with the atmospheric temperature (○). The cylindrical photo-bactericidal apparatus was operated from October 1 to October 21, 2002. Conditions: catalyst, Sb(tpp)/SiO_2 (4 kg); water, ca. 800 L; flow rate, 28 L min^{-1}; and average retention time, 26 s.

leaf-type of photoinactivation apparatus (200 mmϕ × 50 mm, **Figure 11B**) containing the Sb(tpp)/SiO_2 catalyst (80 g, 0.05 wt% Sb(tpp)), which operated under sunlight irradiation. The determination of viable cell numbers of *Legionella* species was carried out in the manner described in Section 4.2. After the leaf-type apparatus had been installed into the fountain, the concentrations of *Legionella* species in the fountain were continuously kept below 30 CFU under sunlight irradiation (**Figure 12**). After the leaf-type apparatus was removed, the concentrations of *Legionella* species gradually increased to reach 100 CFU, which is the environmental quality standard, within 42 days after the removal.

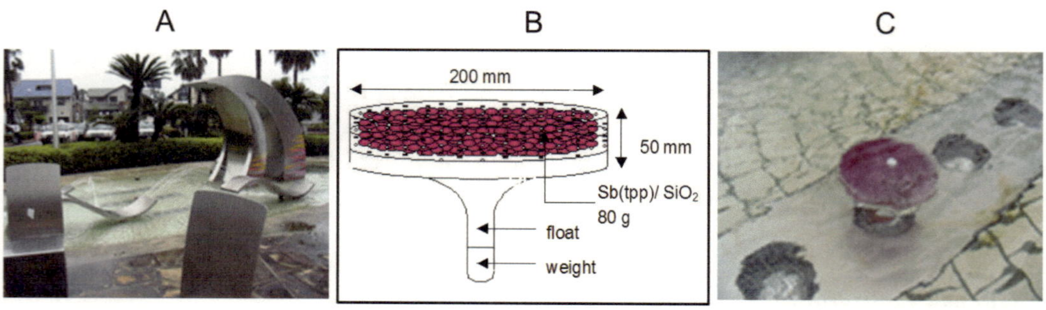

Figure 11. Practical experiments in a public fountain: The leaf-type apparatus (B) was set in the fountain (A). (C) Picture indicates the leaf-type apparatus set in the pool of the fountain.

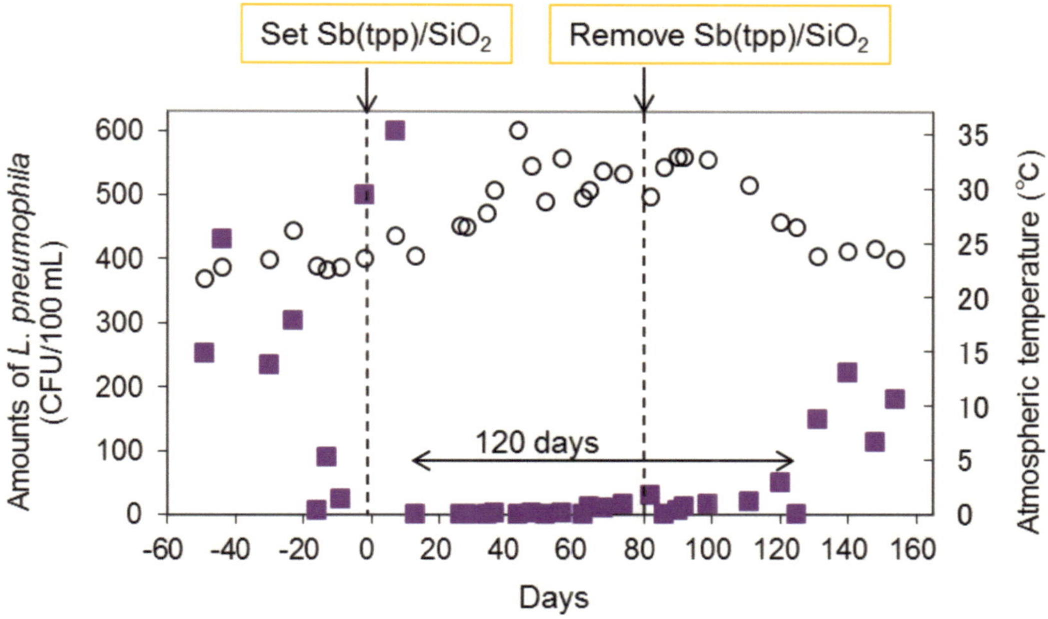

Figure 12. Time-conversion plots of the amount of *Legionella* species (■) in public fountain photoinactivation. The experiment was performed using a leaf-type photo-bactericidal apparatus containing Sb(tpp)/SiO_2 (80 g) in a fountain that contained 13 m^3 of water between April 9 to October 29, 2003. The leaf-type apparatus was installed into the fountain on May 28, 2003 and was removed from the fountain on August 22, 2003. Atmospheric temperature (○) was recorded as the average of the highest temperature of Miyazaki city for three days before the sampling day.

As a result of the practical experiments, it is noteworthy that 80 g of Sb(tpp)/SiO_2 catalyst, which contained 40 mg of Sb(tpp), could maintain the concentration of *Legionella* species in 13 m^3 of water below 100 CFU for 120 days.

4.4. Mechanism for photoinactivation using M(tpp)/SiO_2

Elemental analyses of the catalysts before and after use in the fountain were performed with ICP. Before use, the Sb content in Sb(tpp)/SiO_2 was measured to be 80 ppm, which was in good agreement with the Sb content (72 ppm) calculated for the 0.05 wt% of Sb(tpp) content in the catalyst. After 3 months of use in the fountain, the Sb content decreased from 80 to 17 ppm. On the other hand, Na, Mg, Al, and Ca largely increased, resulting in ion-absorption on SiO_2. Moreover, Sb(tpp)/SiO_2 catalyst used in the fountain was analyzed by a confocal laser scanning microscopy (CLSM). It was found that the fluorescence coming from the surface of the catalyst keep the similar shapes to the original catalyst, but the intensity was weaker compared with the original spectra of Sb(tpp)/SiO_2. On the other hand, the fluorescence from the inside of the catalyst maintained the original intensity. Therefore, it is suggested that Sb(tpp) was eliminated from the surface of the catalyst. Irradiation of fluorescent light on the Sb(tpp)/SiO_2 catalyst in deionized water did not sufficiently account for the spectral change and decrease in total Sb. Therefore, that the cationic Sb(tpp) chromophore was exchanged with alkali metal ions in the bulk water on the surface of the catalyst under irradiation is strongly suggested.

A similar phenomena were observed in the case of P(tpp)/SiO_2 [10]. Moreover, when Sb(tpp) was tightly fixed on SiO_2 through covalent bonds, no photoinactivation occurred [19].

Therefore, the liberation of the Sb(tpp) chromophore from SiO_2 is necessary for photoinactivation, as shown in **Figure 13**. Sb(tpp) can dissolve slightly in water (the water solubility (C_W) of Sb(tpp) is low (0.08 mM)). The liberated Sb(tpp) might be adsorbed by the bacteria and induce apoptosis under visible light irradiation. After the photoinactivation, Sb(tpp) might separate from the SiO_2 and move to the pool of the fountain. This may explain the loss of Sb(tpp) from Sb(tpp)/SiO_2.

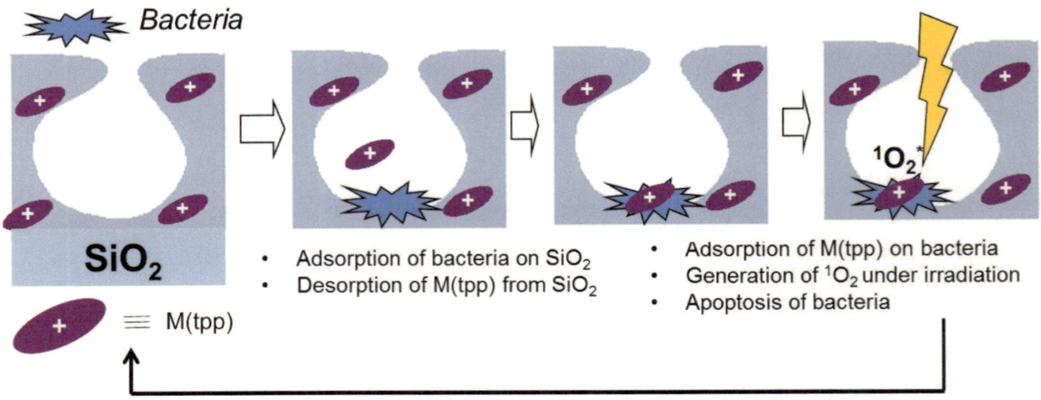

Figure 13. Possible mechanism for photoinactivation of bacteria by M(tpp)/SiO_2.

5. Photosensitized inactivation using water-soluble porphyrins

As mentioned in the previous section, it was found that Sb(tpp) dissolved in water was responsible for the photoinactivation of bacteria. Therefore, in our next study, we intended to inactivate bacteria using water-soluble porphyrins. Water-soluble porphyrins have received much attention in connection with photoinactivation [20] and photodynamic therapy (PDT) [21–24] ever since the first report on photoinactivation of *E. coli* by water-soluble *meso*-substituted cationic porphyrins appeared in 1996 [25]. For the biological application of porphyrins, water solubility is an important characteristic in the handling of porphyrins in an aqueous solution. We modified the axial ligands of P and Sb porphyrins by installing alkyloxo (**1**) [26], alkylethylenedioxy (**2**) [27, 28], and alkylpyridinium groups (**3**) [20, 29] (**Figure 14**). The water solubility (C_W in mM) is shown in **Table 2**. The quantum yields for the formation of $^1O_2$ were determined to be 0.65 for **1b**, 0.53 for **1d** [26], 0.62 for **2b**, 0.69 for **2c**, 0.73 for **2d** [27], 0.88 for **3g**, and 0.87 for **3c** [29].

We show the results of photoinactivation of *S. cerevisiae* and *E. coli* using water-soluble P and Sb porphyrins (**1–3**). Photoinactivation was enacted using the apparatus shown in **Figure 7**. The porphyrin solution, a bacteria suspension, and buffer solution (or water) were introduced into L-type glass tubes, resulting in a solution (10 mL) containing bacteria cells

http://dx.doi.org/10.5772/intechopen.70004

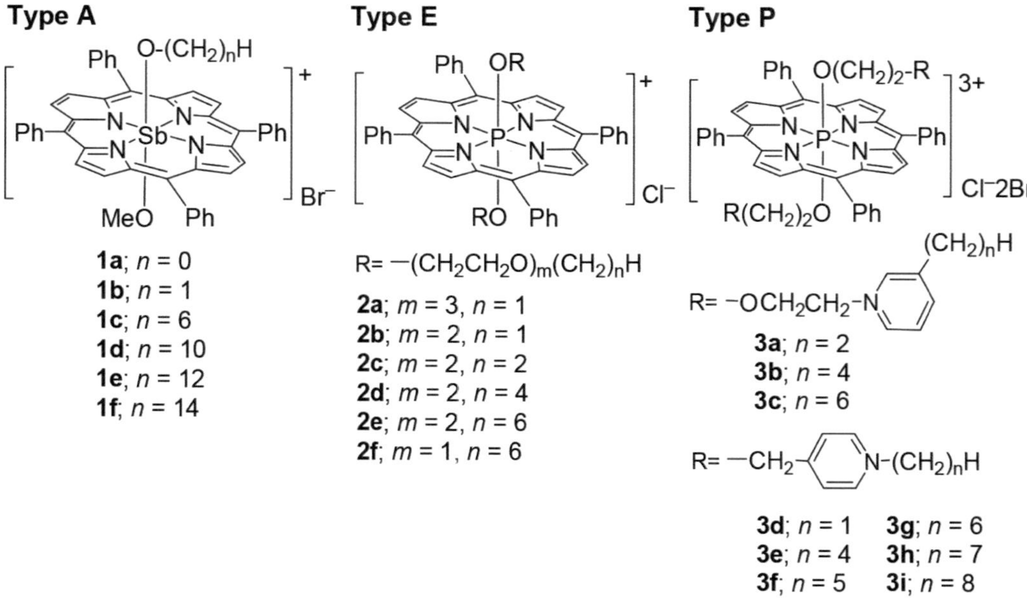

Figure 14. Water-soluble porphyrins (**1–3**).

(ca. 1 × 10^4 cell mL^{-1}) and compounds **1–3** (5–2000 nM). The solution was irradiated using a fluorescent lamp (λ_{max} = 545 nm) under aerobic conditions on a reciprocal shaker. The amounts of bacteria (cell mL^{-1}) at every irradiation time were determined by the colony counting method. Under dark conditions, the bacterial counts were maintained at the initial levels. Thus, it was confirmed that all porphyrins (**1–3**) have no bactericidal activity under dark conditions at all.

From the plots of the survival ratios against irradiation time, the bactericidal activity of porphyrins was evaluated by the half life ($T_{1/2}$ in min), which was the time required for the initial bacteria concentration to be halved. Moreover, the minimum concentrations ([P]) of porphyrins were adjusted until $T_{1/2}$ was found in the range of 0–120 min. The [P] values and $T_{1/2}$ are shown in **Table 2**. As [P] values and $T_{1/2}$ are smaller, the bactericidal activity of porphyrins is higher. The most active sensitizer for *S. cerevisiae* was **2e** and **2f**, whose [P] value was 5.0 nM. However, **2e** and **2f** were ineffective for *E. coli*. Tricationic complexes **3a–3c** were effective for *E. coli*. The bioaffinity of the porphyrins can be related to the structure of the bacterial cell wall. In the case of *S. cerevisiae*, whose cell wall consisted of hydrophobic peptidoglycan, water-soluble porphyrins (**2e–2f**) having hydrophobic character had the highest bioaffinity. On the other hand, polycationic porphyrins (**3**) had the highest bioaffinity toward gram-negative *E. coli*, whose cell wall consisted of phospholipids, lipopolysaccharides, lipoteichoic acids, and lipoproteins [30]. Moreover, it was found that **3** interacted strongly with human serum albumin (HSA) [31], and the addition of HSA was effective for inducing photoinactivation of *S. cerevisiae* using **3** [29]. Recently, reviews on photoinactivation by water-soluble porphyrins have been published by Almeida at al. [20] and our group [32].

Entry	*n*	*m*	C_w/mM	*S. cerevisiae*[b]		*E. coli*	
				[*P*]/nM	$T_{1/2}$/min	[*P*]/nM	$T_{1/2}$/min
1a	0		0.10	50	380	–	–
1b	1		0.13	50	192	–	–
1c	6		1.09	50	14	–	–
1d	10		2.10	40	22	–	–
1e	12		2.21	50	17	–	–
1f	14		2.40	50	21	–	–
2a	1	3	17.4	500	23	–	–
2b	1	2	13.9	300	81	–	–
2c	2	2	13.0	200	31	–	–
2d	4	2	5.38	50	55	–	–
2e	6	2	2.07	5	64	–	–
2f	6	1	1.11	5	69	–	–
3a	2		>120	*50	*32	250[c]	32[c]
3b	4		112	*50	*36	250[c]	53[c]
3c	6		63.6	50	44	250[c]	120[c]
3d	1		3.35	50	20	2000	66
3e	4		6.10	*30	*85	2000	27
3f	5		3.80	–	–	500	29
3g	6		5.84	*20	*72	500	31
3h	7		6.00	–	–	400	24
3i	8		3.80	–	–	500	63

[a]Water solubility of porphyrins in mM.
[b]Experiment with * was performed in the presence of human serum albumin (HSA, 400 nM).
[c]Unpublished results.

Table 2. Photoinactivation of *S. cerevisiae* and *E. coli* by water-soluble porphyrins (**1**, **2**, and **3**).

6. Conclusion and perspectives

Since aqueous solutions are more transparent for visible light than ultraviolet, visible light photocatalysts work best for the photocatalytic reactions in aqueous solution. Moreover, visible light photocatalysts take advantage of the photocatalytic reactions under sunlight irradiation, since sunlight consists of 52% visible, 42% infrared, 6% UV-A, and 0.5% UV-B light. We showed two methods to photoinactivate bacteria: one method is the dispersion of M(tpp)/SiO_2 (M = P, Ge, and Sb) in water, which is applicable to open system in naturally occurring environments; the other method is the water solubilization of M-porphyrins (M = P and Sb), which can be used in a closed system. M(tpp) and M-porphyrins can interact with bacteria through adsorption onto cell walls and absorption into the cells. Under irradiation, reactive species, such as $^1O_2$, is generated by energy transfer from the porphyrins to O_2 molecules on the cell walls and inside the cells.

Thus, porphyrins are useful chromophores for catalysis and sensitization in a biological application. The application of water-soluble M-porphyrins to PDT is currently underway in our laboratories.

Acknowledgements

This work was partially supported by a Grant-in-Aid for Scientific Research (C) (16K05847) from Japan Society for the Promotion of Science (JSPS). Also, the authors would like to express grateful acknowledgment to many students belonged in our laboratory during 2003 to present when visible light inactivation has been investigated in our laboratory for their large contributions.

Author details

Masahide Yasuda* and Jin Matsumoto

*Address all correspondence to: yasuda@cc.miyazaki-u.ac.jp

Department of Applied Chemistry, Faculty of Engineering, University of Miyazaki, Miyazaki, Japan

References

[1] Fujishima A, Rao TN, Tryk DA. Titanium dioxide photocatalysis. Journal of Photochemistry and Photobiology C: Photochemistry Reviews. 2000;**1**:1-21

[2] Matsunaga T, Tomoda R, Nakajima T, Wake H. Photoelectrochemical sterilization of microbial cells by semiconductor powders. FEMS Microbiology Letters. 1985;**29**:211-214

[3] Fujitsuka M, Cho DW, Tojo S, Inoue A, Shiragami T, Yasuda M, Majima T. Electron transfer from axial ligand to S1- and S2-excited phosphorus tetraphenylporphyrin. The Journal of Physical Chemistry A. 2007;**111**:10574-10579

[4] Shiragami T, Shiraki R, Makise R, Matsumoto J, Yasuda M. Di(hydroxo)porphyrin Ge^{IV} complex/silica gel composite as visible light-assisted radical generator. Chemistry Letters. 2010;**39**:874-875

[5] Fujitsuka M, Shiragami T, Cho DW, Tojo S, Yasuda M, Majima T. Solvent dynamics regulated electron transfer in S2-excited Sb and Ge tetraphenylporphyrins with an electron donor substituent at the *meso*-position. The Journal of Physical Chemistry A. 2014;**118**:3926-3933

[6] Tsunami SI, Tanaka K, Matsumoto J, Shiragami T, Yasuda M. Effects of an axial amino ligand on the spectroscopic and electrochemical properties of amino(methoxo)-(tetraphenylporphyrinato)antimony(V) complexes. Bulletin of the Chemical Society of Japan. 2008;**81**:583-589

[7] Andou Y, Shiragami T, Shima K, Yasuda M. Synthesis and fluorescence study of a (2-naphthoxy)polyoxalkoxyantimony(V) tetraphenylporphyrin complex. Journal of Photochemistry and Photobiology A: Chemistry. 2002;**147**:191-197

[8] Felton RH, Linschitz H. Polarographic reduction of porphyrins and electron spin resonance of porphyrin anions. Journal of the American Chemical Society. 1966;**88**:1113-1116

[9] Shiragami T, Matsumoto J, Inoue H, Yasuda M. Antimony porphyrin complexes as visible-light driven photocatalyst. Journal of Photochemistry and Photobiology C: Photochemistry Reviews. 2005;**6**:227-248

[10] Fueda Y, Suzuki H, Komiya Y, Asakura Y, Shiragami T, Matsumoto J, Yokoi H, Yasuda M. Bactericidal effect of silica gel-supported porphyrinatophosphorus(V) catalysts on *Escherichia coli* under visible-light irradiation. Bulletin of the Chemical Society of Japan. 2006;**79**:1420-1425

[11] Yokoi H, Shiragami T, Hirose J, Kawauchi T, Hinoue K, Fueda Y, Nobuhara K, Akazaki I, Yasuda M. Bactericidal effect of a silica gel-supported porphyrinatoantimony(V) complex under visible light irradiation. World Journal of Microbiology and Biotechnology. 2003;**19**:559-563

[12] Shiragami T, Makise RI, Inokuchi Y, Matsumoto J, Inoue H, Yasuda M. Efficient photocatalytic oxidation of cycloalkenes by dihydroxo(tetraphenylporphyrunato) antimony supported on silica gel under visible light irradiation. Chemistry Letters. 2004;**33**:736-737

[13] Shiragami T, Shimizu Y, Hinoue K, Fueta Y, Nobuhara K, Akazaki I, Yasuda M. Silica gel-supported porphyrinatoantimony(V) complex acting as visible-light driven photocatalyst for dechlorination of chlorophenols in aqueous solution. Journal of Photochemistry and Photobiology A: Chemistry. 2003;**156**:115-119

[14] Chang FY, Yu VL. *Legionella* infection. In: Fishman AP, Elias JA, Fishman JA, Grippi MA, Kaiser LR, and Senior R, editors. Fishman's Pulmonary Diseases and Disorders. Vol. 2, 3rd ed. New York: McGraw Hill; 1996. pp. 945-949

[15] Fernández JA, López P, Orozco D, Merino J. Clinical study of an outbreak of Legionnaire's disease in Alcoy, southeastern Spain. European Journal of Clinical Microbiology & Infectious Diseases. 2002;**21**:729-735

[16] Lettinga KD, Verbon A, Weverling GJ, Schellekens JFP, Den Boer JW, Yzerman EPF, Prins J, Boersma WG, Van Ketel RJ, Prins JM, Speelman P. Legionnaires' disease at a Dutch flower show: Prognostic factors and impact of therapy. Emerging Infectious Diseases Journal. 2002;**8**:1448-1454

[17] Correia AM, Gonçalves G, Reis J, Cruz JM, Castro e Freitas JA. An outbreak of legionnaires' disease in a municipality in northern Portugal. Euro Surveillance. 2001;**6**:121-124

[18] Fueda Y, Hashimoto M, Nobuhara K, Yokoi H, Komiya Y, Shiragami T, Matsumoto J, Kawano K, Suzuki S, Yasuda M. Visible-light bactericidal effect of silica gel-supported porphyrinatoantimony(V) catalyst on *Legionella* species occurring in the living environmental fields. Biocontrol Science. 2005;**10**:55-60

[19] Matsumoto J, Matsumoto T, Senda Y, Shiragami T, Yasuda M. Preparation and characterization of porphyrin chromophores immobilized on micro-silica gel beads. Journal of Photochemistry and Photobiology A: Chemistry. 2008;**197**:101-109

[20] Alves E, Faustino MAF, Neves MGPMS, Cunha T, Nadais H, Almeida A. Potential applications of porphyrins in photodynamic inactivation beyond the medical scope. Journal of Photochemistry and Photobiology C: Photochemistry Reviews. 2014;**22**:34-57

[21] Josefsen LB, Boyle RW. Unique diagnostic and therapeutic roles of porphyrins and phthalocyanines in photodynamic therapy, imaging and theranostics. Theranostics. 2012;**2**:916-966

[22] Kudinova NV, Berezov TT. Photodynamic therapy of cancer: Search for ideal photosensitizer. Biochemistry (Moscow) Supplement Series B Biomedical Chemistry. 2010;**4**:95-103

[23] Nyman ES, Hynninen PH. Research advances in the use of tetrapyrrolic photosensitizers for photodynamic therapy. Journal of Photochemistry and Photobiology B: Biology. 2004;**73**:1-28

[24] Wainwright M. Photodynamic therapy: The development of new photosensitisers. Anti-Cancer Agents in Medicinal Chemistry. 2008;**8**:280-291

[25] Merchat M, Spikes JD, Bertoloni G, Jori G. Studies on the mechanism of bacteria photosensitization by *meso*-substituted cationic porphyrins. Journal of Photochemistry and Photobiology B: Biology. 1996;**35**:149-157

[26] Yasuda M, Nakahara T, Matsumoto T, Shiragami T, Matsumoto J, Yokoi H, Hirano T, Hirakawa K. Visible light-assisted sterilization activity of water-soluble antimonyporphyrin toward *Saccharomyces cerevisiae*. Journal of Photochemistry and Photobiology A: Chemistry. 2009;**205**:210-214

[27] Matsumoto J, Shinbara T, Tanimura SI, Matsumoto T, Shiragami T, Yokoi H, Nosaka Y, Okazaki S, Hirakawa K, Yasuda M. Water-soluble phosphorus porphyrins with high activity for visible light-assisted inactivation of *Saccharomyces cerevisiae*. Journal of Photochemistry and Photobiology A: Chemistry. 2011;**218**:178-184

[28] Yasuda M, Shiragami T, Matsumoto J. Water-soluble Porphyrin and Process for Production Thereof. JP Patent JP20090233246; 2009

[29] Matsumoto J, Kai Y, Yokoi H, Okazaki S, Yasuda M. Assistance of human serum albumin to photo-sensitized inactivation of *Saccharomyces cerevisiae* with axially pyridinio-bonded P-porphyrins. Journal of Photochemistry and Photobiology B: Biology. 2016;**161**:279-283

[30] Matsumoto J, Suemoto Y, Kanemaru H, Takemori K, Shigehara M, Miyamoto A, Yokoi H, Yasuda M. Alkyl substituent effect on photosensitized inactivation of *Escherichia coli* by pyridinium-bonded P-porphyrins. Journal of Photochemistry and Photobiology B: Biology. 2017;**168**:124-131

[31] Matsumoto J, Kubo T, Shinbara T, Matsuda N, Shiragami T, Fujitsuka M, Majima T, Yasuda M. Spectroscopic analysis of the interaction of human serum albumin with tricationic phosphorus porphyrins bearing axial pyridinio groups. Bulletin of the Chemical Society of Japan. 2013;**86**:1240-1247

[32] Matsumoto J, Shiragami T, Hirakawa K, Yasuda M. Water-solubilization of P(V) and Sb(V) porphyrins and their photobiological application. International Journal of Photoenergy. 2015;**2015**:Article ID 148964

Chapter 4

Carbon Quantum Dots: A Component of Efficient Visible Light Photocatalysts

Shan Cong and Zhigang Zhao

Additional information is available at the end of the chapter

http://dx.doi.org/10.5772/intechopen.70801

Abstract

Carbon quantum dots (CQDs) have been developed as a new member of nanocarbons, characterized by the relatively easy preparation from a wide spectrum of carbonaceous precursors through either bottom-up or top-down routes. Attractive optoelectronic properties have been observed with CQDs, including efficient light absorption, variable photoluminescence (PL), unique up-conversion PL and prominent electron transport ability, which make CQDs an important component with great potential in the design of efficient visible light-driven photocatalysts. In this chapter, detailed contribution of CQDs to the enhanced visible light-driven photocatalysis will be included, in the classification of the role as electron mediator, photosensitizer, spectral converter and sole photocatalyst.

Keywords: carbon quantum dots, photocatalysis, visible light, charge carrier, photoluminescence

1. Introduction

Carbon dots have emerged as a new class of quantum dot-like nanocarbons, which are typically constituted by discrete, quasi-spherical nanoparticles with sizes below 10 nm. The first discovery of quantum dot-like nanocarbons can be dated back to 2004, when carbon nanoparticles with sizes at about 1 nm were isolated by Xu et al. [1] as a byproduct of preparation of single-walled carbon nanotubes, separated from carbon soot produced by arc discharge. However, it was until 2 years later, these new kind of nanocarbons have been documented as "carbon quantum dots" (CQDs) and boosted widespread interests, since the synthesis of fluorescent carbon nanoparticles with diameter less than 10 nm from Sun's group [2].

Distinct from other well developed nanocarbons, such as fullerenes, carbon nanotubes and graphene, CQDs show their unique advantages, including isotropic shapes, ultrafine dimensions, tunable surface functionalities as well as the simple, fast and cheap preparations, which are attractive for a host of applications. Meanwhile, as a new member of quantum dot materials, CQDs also show potential as replacement for traditional toxic metal-based quantum dots currently in use, benefiting from their highly hydrophilic surface, high resistance to photobleaching, easy functionalization, chemical inertness, low toxicity and good biocompatibility. Until now, the applications of CQDs have been widely demonstrated in several fields, ranging from sensing, bioimaging to catalysis and energy conversion/storage.

This chapter will start with a brief overview about the preparation routes of CQDs, followed by a quick snapshot of their crystalline and electronic structure as well as the structure-induced optical properties. Then, the role of CQDs as an active component in modern visible light photocatalysis will be discussed in detail, based on a short review about works published in recent years.

2. Preparation methods

The preparation of CQDs can be generally classified into "top-down" and "bottom-up" routes (**Figure 1**). The former involves breaking down of carbonaceous materials via chemical, electrochemical or physical approaches. On the contrary, the latter is realized by carbonization of small organic molecules followed by chemical fusion. A detailed review about the preparation of CQDs can be found in several review articles [3–9], and only a brief overview of recent progresses about the CQDs preparation will be given here.

2.1. Top-down route

The first reported CQDs that isolated by chance from arc-discharged soot were produced by the top-down method, in which the carbonaceous soot was cleaved through nitric acid

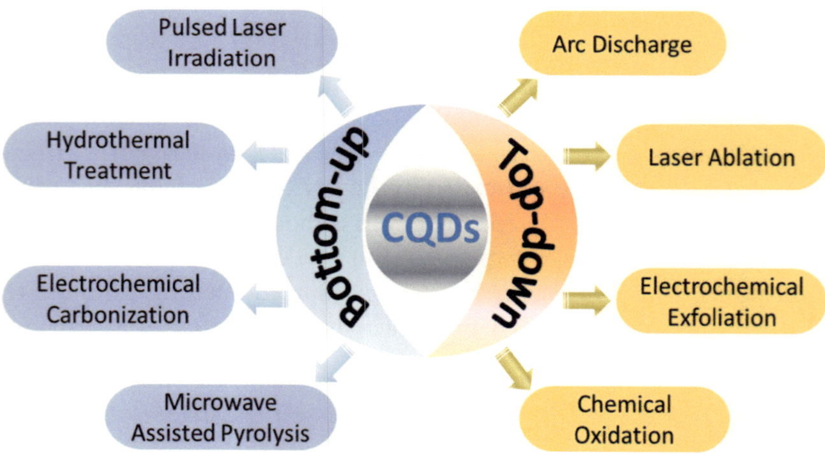

Figure 1. Schematic illustration of CQD preparation via "top-down" and"bottom-up" approaches.

oxidation, followed by gel electrophoresis separation to extract CQDs [1]. Subsequently, laser ablation of graphite using Ar as a carrier gas in the presence of water vapor was proved to be effective for the preparation of CQDs by means of the top-down route [2]. To date, CQDs have already been produced from graphite [10], graphene [11], carbon nanotubes (CNTs) [12] and other materials with sp2 carbon structure as a product of dimension reduction, meanwhile, through various physicochemical processes, including arc discharge [1], laser ablation [2], electrochemical exfoliation [13, 14] and chemical oxidation [15, 16]. However, to break down the chemical inert carbon structure of sp2 covalence bonds, the top-down process is invariably complicated with harsh conditions, some may be environmentally harmful, and thus not suitable for a large-scale production. In addition, effective control over the particle size of CQDs product is often difficult to realize in such top-down processes.

2.2. Bottom-up route

On the contrary, the bottom-up methods offer opportunities to control the growth of CQDs by using organic molecular as carbon precursors. Carbonization reaction can be applied to various kinds of molecules, followed by aggregation in solution to produce CQDs. Pulsed laser irradiation of toluene [17], hydrothermal treatment of citric acid [18], electrochemical carbonization of low-molecular-weight alcohols [19] and microwave-assisted pyrolysis of citric acid formamide solution [20] have already been utilized for the preparation of CQDs. Recently, biomass molecules, such as sucrose [21], glucose [22], cellulose [23] and amino acid [24], have attracted great attentions as suitable precursors for the preparation of CQDs via dehydrate and further carbonize. Moreover, raw biomass is also suitable precursor for the preparation of CQDs, as a strategy potential for large-scale production. For instance, Sahu et al. [25] prepared 0.4 g of CQDs from a mixture of orange juice with ethanol by hydrothermal treatment at 120°C for 2.5 h (**Figure 2A**). While using chicken eggs as raw material, Chen and coworkers [26] prepared CQDs via plasma-induced pyrolysis with a yield up to 10 g, suitable for applications as fluorescent inks (**Figure 2B**). In comparison with the liquid biomass as precursor, solid biomass may provide a more practical solution to scale up the synthesis of CQDs due to their much longer shelf life. For example, hair can be used as carbon source for the synthesis of CQDs, in utilization of its main component of keratin. A one-step thermal treatment of hair at 200°C for about 24 h was reported by Guo et al. to obtain CQDs, with the yield estimated to

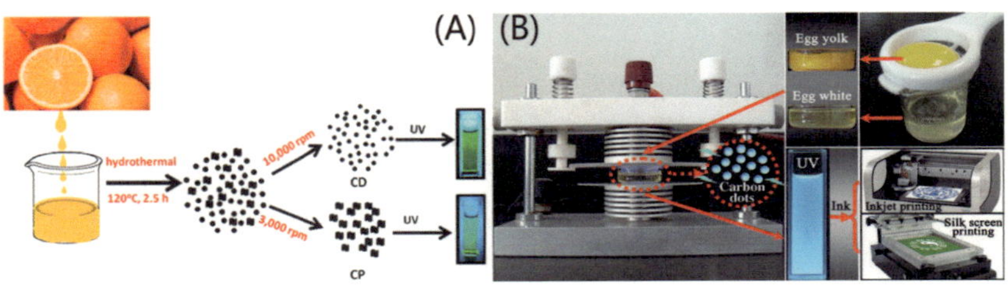

Figure 2. (A) Illustration of the formation of CQDs from orange juice by hydrothermal treatment. (B) Fabrication of CQDs from chicken egg and their application as fluorescent carbon inks. Reprinted with permission [25, 26]. Copyright 2012, Royal Society of Chemistry; Copyright 2012, Wiley-VCH.

be about 95% [27]. Thanks to the diversity and plenty of biomass, the large-scale synthesis of CQDs from biomass is of great potential especially in view of practical applications.

3. Structures and properties of CQDs

3.1. Structures

Generally, CQDs are quasi-spherical nanoparticles with diameter less than 10 nm, which can be amorphous or crystalline nanoparticles built up with sp2 carbon clusters. A typical fringe spacing of 0.34 nm can readily be distinguished from the high resolution transmission electron microscopy (HR-TEM) images of well crystalized CQDs, corresponding to the (002) interlayer spacing of graphite (**Figure 3A**) [28]. While as a special example of CQDs, graphene quantum dots (G-CQDs) usually possess a good crystallinity and layered structure (composed by less than 4, mostly a single graphene layer), with a crystal lattice spacing of 0.24 nm, corresponding to the (100) in-plane lattice parameter of graphene (**Figure 3B**) [29].

The modulated electronic structure of CQDs is unique among carbonaceous materials, which is dramatically size dependent. According to the quantum confinement effect (QCE), a reduced band gap of CQDs can be expected when the dot size increases. Li et al. [30] performed theoretical calculations to investigate the relationship between luminescence and cluster size of CQDs, who confirmed that the band gap reduced gradually as the size of CQDs increased (**Figure 3C**). Meanwhile, variable functional groups (such as hydroxyl and carboxyl) anchored on the surface of CQDs are indispensable for their excellent aqueous solubility, which can also greatly affect the electronic structure of CQDs. It has been demonstrated that different functional groups (either electron withdrawing or electron donating) simultaneously modulate the lowest unoccupied molecular orbital (LUMO) and highest occupied molecular orbital (HOMO) levels of CQDs, however, with relative small change in the band gap [31]. For instance, an electron donating group could raise both the HOMO and LUMO levels to higher energy, while an electron withdrawing group would cause a prominent down-shift of the total energy levels (**Figure 3D**). In addition, doping with heteroatoms such as nitrogen, sulfur were found to alter the electronic structures of CQDs significantly. The incorporation of ~1 at% of S into CQDs could effectively modify their electronic structure by introducing S-related energy levels between π and π^* orbitals of CQDs, which would diversify the electron transition pathways and facilitate inter band crossing of electrons (**Figure 3E**) [32].

Overall, the diversity of modulations for CQDs, from size control to surface functional groups exchange and heteroatom doping, makes it possible in the delicate control over their electronic structures and optical properties, showing potential as active components in light harvesting and utilizations.

3.2. Optical properties

Most notable, unique optical properties are discovered with CQDs, including wide-spectrum light harvesting from ultraviolet to near-infrared (NIR), tunable photoluminescence (PL) and

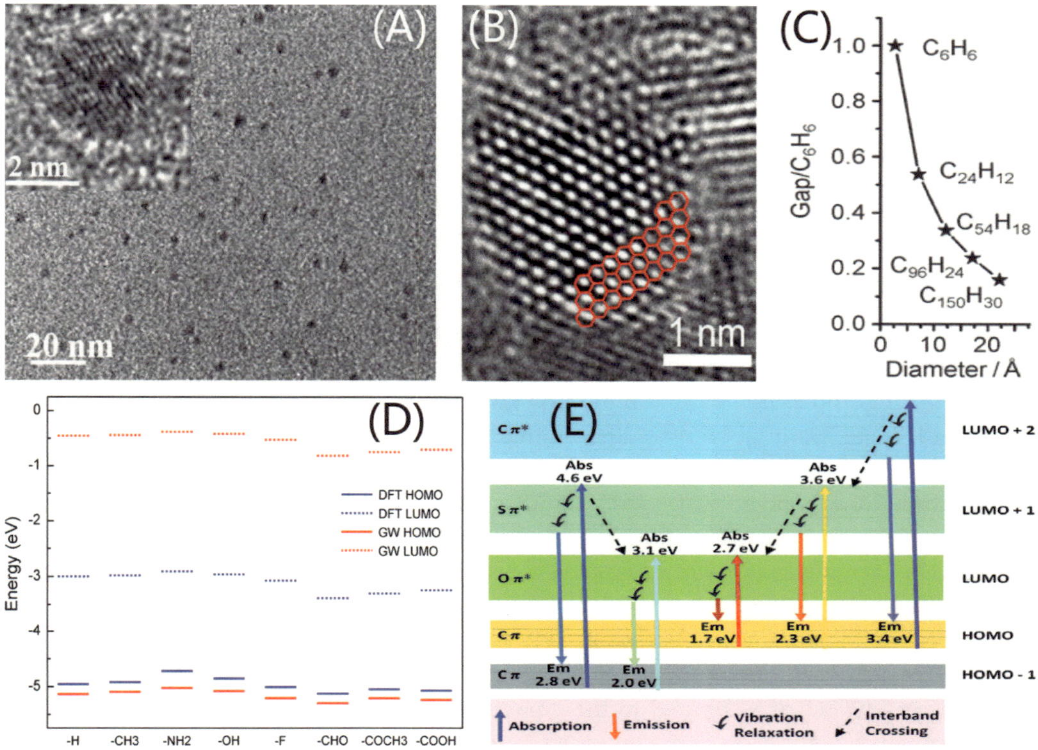

Figure 3. (A) TEM images of CQDs. (B) HR-TEM image of sp2 domains G-CQDs. (C) LUMO-HOMO gap dependence on the size of CQDs. (D) Energy diagrams of CQDs functionalized with different groups. (E) The proposed energy level diagram of S-doped CQDs. Reprinted with permission [28–32]. Copyright 2011, Wiley-VCH; copyright 2015, Royal Society of Chemistry; copyright 2010, Wiley-VCH; copyright 2015, American Chemical Society; copyright 2014, Royal Society of Chemistry.

efficient multiphoton up-conversion, all of which can be effectively tailored by the artificial control of the size, shape, surface and doping, based on the remarkable quantum confinement effect (QCE), surface effect and edge effect.

3.2.1. *Absorbance*

CQDs are effective photon-harvesting agents especially in short-wavelength region, due to the π-π* transition of C═C bonds. Typically, CQDs show strong optical absorption in the UV region (260–320 nm) with a tail extending into the visible range, in which a maximum peak around 230 nm is assigned to the π-π* transition of aromatic C═C bonds and a shoulder at 300 nm is assigned to the n-π* transition of C═O bonds or other surface groups [33]. In addition, the surface functional groups may also greatly contribute to the absorption at the UV-visible region, since a gradual red-shifted absorption can be observed after modification with some passivating agents. For instance, the absorbance of CQDs was found to increase to longer wavelength in the range of 350–550 nm after surface passivation with 4,7,10-trioxa-1,13-tridecanediamine (TTDDA) [34]. Moreover, doping with heteroatom is also effective to tailor the absorption of CQDs. As reported by Qu et al. [35], the absorption band of S,N-co-doped CQDs moved into the visible region of 550–595 nm, showing potential as visible light-harvesting agent.

3.2.2. Photoluminescence (PL)

Tunable photoluminescence, with an emission peak ranging from deep ultraviolet to visible or even NIR region, is one of the most fascinating features of CQDs. Usually, these observed tunable PL properties are in close relationship with the QCE or surface effects of CQDs.

It is wildly accepted that the PL of CQDs originates from the emissive surface energy traps upon stabilization, since surface passivation is often indispensable for photoluminescent CQDs with high quantum yield (**Figure 4A**) [36]. Meanwhile, some researchers suggested that the PL in CQDs may be a result of radiative recombination of surface-confined electrons and holes, based on the observation that PL of CQDs can be effectively quenched by both electron donors and acceptors [37]. Significantly, this notion indicates that photo-excited CQDs are both good electron acceptors and electron donors. While in some cases, quantum size effect is the main reason for the observed size-dependent optical properties of CQDs. As reported by Li et al. [30], CQDs emitted blue, green, yellow, and orange, respectively, along with increased dot size under identical λ_{ex} (**Figure 4B**). Since no appreciable change in the PL spectra after further H_2 plasma treatment to remove surface oxygen species, the luminescence emissions of CQDs was attributed to the quantum-sized graphite structure rather than the carbon-oxygen surface. However, contradict results also existed, as Bao et al. [28] observed a red-shifted PL emission peak when the sizes of CQDs decreased at the same λ_{ex}. Based on the assumption that smaller CQDs were easily oxidized to take on more oxygen-related surface states, they suggested the red-shifted PL of CQDs was a result of variations in surface states rather than a size effect.

Therefore, the origin of the displayed optical properties of CQDs is still in debate, which may be probably in relationship with the competition between the emissive sites and non-radiative trap sites on the surface, as well as the quantum confinement effects determined by particle sizes.

3.2.3. Up-conversion photoluminescence (UCPL)

In contrary to the conventional down-conversion fluorescence emissions (Stokes PL), certain CQDs are also found to display up-conversion fluorescence emissions (anti-Stokes PL), with the emission wavelength shorter than the excitation wavelength, arising from the simultaneous absorption of two or multiple longer wavelength photons. UCPL is an attractive optical property of CQDs, which enables many promising applications, especially for photocatalysis in the visible to NIR spectrum.

The visible emission of ~5 nm CQDs under the excitation of 800 nm femtosecond pulsed laser was first recorded by Cao et al. [38], who proposed a two-photon mechanism for the up-conversion emission. Later, several other groups observed similar UCPL from CQDs prepared via different synthetic routes. For instance, Gan et al. [39] demonstrated that blue light centered

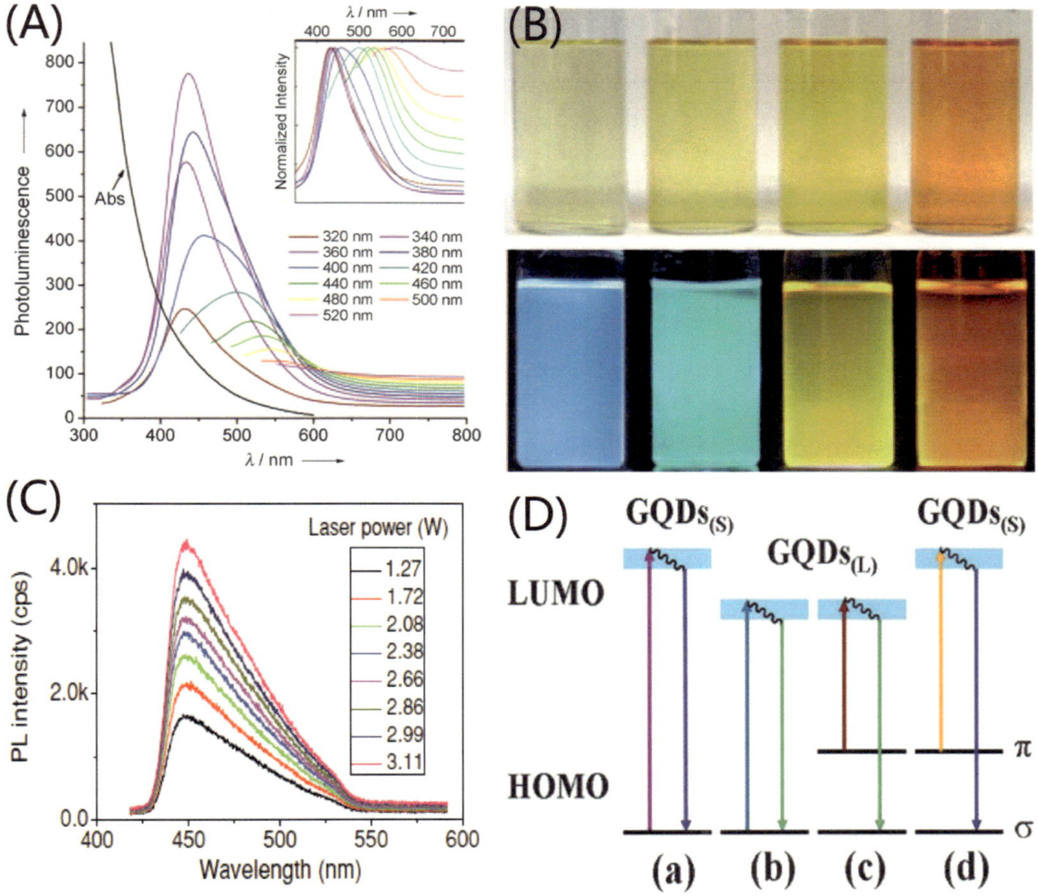

Figure 4. (A) UV/vis absorption (Abs) and PL emission spectra (recorded for progressively longer excitation wavelengths from 320 to 520 nm) of PEG_{1500N} passivated CQDs. (B) Optical images of CQDs illuminated under while (above, daylight lamp) and UV light (bottom, 365 nm). (C) UCPL spectra of CQDs under the excitation of a femtosecond pulsed laser at 800 nm. (D) Schematic illustration of various typical electronic transition processes of CQDs. Normal PL of small size (a) and large size (b); UCPL of large size (c) and large size (d). Reprinted with permission [30, 36, 39]. Copyright 2009, Wiley-VCH; copyright 2010, Wiley-VCH; copyright 2013, Wiley-VCH.

at 450 nm was emitted from CQDs under an excitation of 800 nm femtosecond pulsed laser, and the UCPL intensity was found to relate to the excitation power density (**Figure 4C**).

However, it was still under debate whether the multiphoton-active process was sufficient to explain the UCPL phenomenon of CQDs. In view of the practically constant energy difference of 1.1 eV between the excitation and emission, Shen et al. [40] proposed an anti-stokes transition mechanism using the energy level structural model, who postulated that low energy photons could excite π electrons (in intermediate levels) to LUMO followed by relaxation into σ orbital (HOMO) for the emission of a shorter wavelength photon (**Figure 4D**). Therefore, the constant energy difference here between excitation and emission light is ascribed to the energy gap between π and σ orbitals.

Moreover, Wen et al. [41] recently argued that the observed multiphoton excitation properties in some CQDs might be artificial and could originate from the normal fluorescence, which was excited by the second diffraction leaking in the monochromator of the fluorescence spectrophotometer. Therefore, great cautions should be paid to the observed UCPL of CQDs, and it was suggested that the exact up-conversion from CQDs would be excited under a high enough power density from a femtosecond pulsed laser.

4. Applications in photocatalysis

Based on the generation of charge carriers such as electrons and holes induced by photon absorption, photocatalysis has found great potentials in the application of both environmental remediation and energy conversion, ranging from pollutant degradation to H_2 generation and CO_2 reduction. To generate carriers with sufficient redox potentials, wide band gap semiconductors (such as TiO_2) are often used as photocatalysts, but at the cost of limited utilization of only high energy UV irradiation (~5% of sunlight). At the same time, recombination of charge carriers pervasively exists during their migration through the bulk to the surface in semiconductor particles, where to initiate photocatalytic reactions. Therefore, evolutions of photocatalytic system to overcome the low-usage of sunlight as well as high recombination rate of the photo-induced charge carriers is still an urgent need to improve the overall quantum efficiency for practical applications. Interestingly, all of the optical and structural properties of the CQDs mentioned above meet with the requirement of photocatalytic reactions (such as utilization toward solar spectrum, fast migration of charge carriers and efficient surface redox reactions), indicating the promising application of CQDs in photocatalysis. As a matter of fact, CQDs have shown their importance as a versatile component especially in visible light-driven photocatalysis since their discovery, ascribing to the tunable optical properties (such as absorbance, PL and UCPL) for excellent solar spectrum utilizations, also due to the electron accepting and transporting features of carbon nanomaterials that can direct the flow of photogenerated charge carriers. Therefore, CQDs can act not only as an efficient photocatalyst alone, but also as a multifunctional constituent in the design of photocatalyst for broadened light response and promoted electron-hole separation. In this section, we will summarize the fundamentally multifaceted roles of CQDs as an important component in visible light-driven photocatalysis.

4.1. Electron mediators

Photon-induced charge carriers with a sufficient amount that can migrate to the surface of semiconductors are at the core of photocatalytic reactions. Since charge carriers can easily be traps by various random defects, to collect or mediate the flow of charge carriers for efficient separation is necessary for an effective photocatalyst design. Noble metals with a large work function are commonly deposited onto the surface of semiconductors to act as an electron sink and help in an efficient separation of charge carriers. However, the use of noble metals shows a main drawback of poor economic viability due to their high expense. As a potential

http://dx.doi.org/10.5772/intechopen.70801

electron mediator in placement of noble metals for new design of photocatalysts, CQDs show excellent electron conductivity as well as large electron-storage capacity, benefiting from the inherent sp2 graphitic carbon compositions.

On the one hand, CQDs can act as conductors for efficient charge carrier transport between adjacent semiconductors, due to the electron conductivity from the Ohmic contact at the semiconductor-CQDs interfaces. Miao et al. [42] fabricated a ternary visible light-driven photocatalyst with the construction of g-C_3N_4/CQDs/AgBr, by the in-situ growth of AgBr nanoparticles on CQDs modified g-C_3N_4 nanosheets. The nanocomposite exhibited improved photocatalytic efficiency for RhB degradation under visible irradiation (>420 nm), which was about 4.0, 5.3 and 2.3 times higher than that of AgBr, g-C_3N_4 and g-C_3N_4/AgBr, respectively. Due to the thermodynamically favorable electron transfer upon band alignment, CQDs was believed to act as mediators to promote the electron transfer from g-C_3N_4 to AgBr for an improved separation of electron-hole pairs, responsible for the improved photocatalysis (**Figure 5A**).

On the other hand, CQDs can act as electron sink to accept photogenerated electrons from adjacent semiconductors, and thus to suppress the electron-hole recombination, benefiting from the large electron-storage capacity of carbon nanostructures. Huang et al. [43] obtained CQDs/$ZnFe_2O_4$ photocatalysts via a simple hydrothermal route, which showed eight times larger transient photocurrent response in comparison with that of $ZnFe_2O_4$ alone under visible irradiation ($\lambda > 420$ nm). In a possible photocatalytic mechanism, the authors inferred that CQDs could function as an electron reservoir and transporter over CQDs/$ZnFe_2O_4$ hybrids, to improve both the separation and transfer efficiency of photogenerated charge carriers (**Figure 5B**). Wu et al. [44] demonstrated a case of a bi-quantum dot (QD) photocatalyst for overall water splitting, in which CQDs and $BiVO_4$ QDs were combined together. Without any co-catalysts or sacrificial reagents but under solar light irradiation, H_2 evolution of 0.92 μmol h^{-1} was achieved with such hybrids, about four times that of $BiVO_4$ QDs. In such a two-electron pathway reaction, CQDs and $BiVO_4$ QDs were believed to be reduction and oxidation reaction

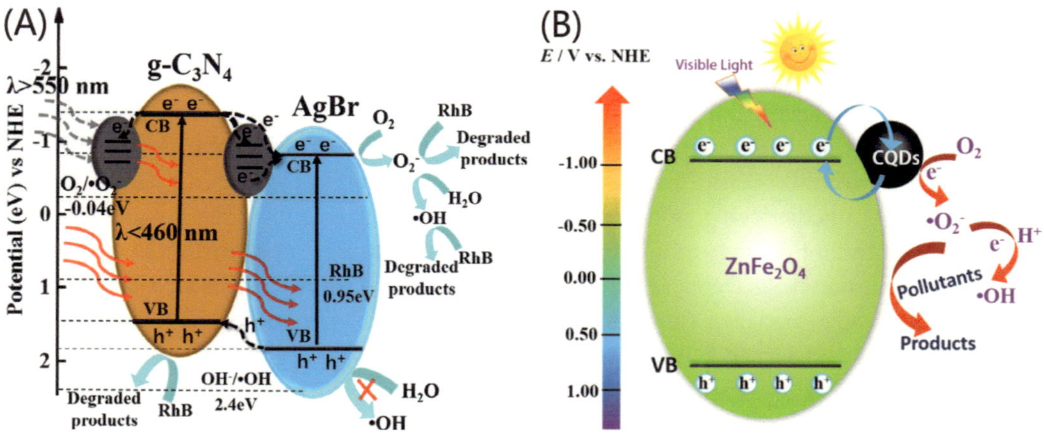

Figure 5. (A) CQDs as electron conductor at the interface of C_3N_4 and AgBr. (B) CQDs as electron sink at the surface of $ZnFe_2O_4$. Reprinted with permission [42, 43]. Copyright 2017, Elsevier; copyright 2017, American Chemical Society.

center, respectively. As an effective electron acceptor, CQDs received electrons from VB of $BiVO_4$ before transferring to surface-adsorbed water for hydrogen production.

Therefore, as electron mediators, CQDs show noble metal comparable ability as both conductor and sink for an efficient charge separation in association with semiconductor photocatalysts. At the same time, the relative low cost and easy preparation make CQDs an available alternative to the expensive noble metal as electron mediators in new design of photocatalysts.

4.2. Photosensitizers

Extending light harvesting, especially by increasing the absorbance of visible photons, has been a popular strategy to achieve improved efficiency for modern photocatalysis. While the extended absorption spectrum of semiconductors usually means narrowed bandgap and thus decreased redox ability, which often obstructs the design of visible light-driven photocatalysts. Using visible light photosensitizers that can transfer photogenerated electrons to the CB of semiconductor in contact has been widely employed to address this conflict. Quantum dots of CdS, CdSe, PbS, are well known photosensitizers with a relative narrow bandgap to achieve good light harvesting efficiency, however, still suffer from some inherent drawbacks, including high cost, instability against photocorrosion and dissolution of toxic heavy-metal ions. Recently discovered CQDs, as an intriguing member of the nanocarbon family, have shown great potentials as photosensitizers due to their broad absorption spectra and large absorption coefficients.

Zhang et al. [45] prepared CQDs with tunable size of 2–5 nm from graphitic polymeric micelles, which showed high water solubility due to the surface decorated carboxyl groups. The obtained CQDs were then loaded onto the surface of P25 TiO_2 through a facial dispersion-evaporation route, followed by thermal treatment at 300°C in air to remove organic passivation layer for a close contact at hetero interfaces. The cooperation of CQDs and TiO_2 was believed to be responsible for the visible light photocatalytic degradation of MB, since either CQDs or TiO_2 alone showed negligible activities. CQDs probably acted as a photosensitizer in the CQDs/TiO_2 hybrid, which injected visible light excited electrons into the CB of TiO_2, responsible for the remarkably enhanced visible light photocatalytic activity (**Figure 6A**). Wu et al. [46] suggested that the surface groups of CQDs were not necessarily removed before using as photosensitizers. CQDs with an average size of 30 nm were prepared through hydrothermal carbonization of pentosan, followed by the preparation of CQDs/TiO_2 composite through directly drying from the suspension of TiO_2 in a CQDs aqueous solution. The CQDs with abundant oxygen-containing groups could act as a photosensitizer to absorb visible light, with excited electrons transferred to the CB of TiO_2. These sensitized electrons then reacted with O_2 adsorbed on TiO_2 surface to produce active oxygen radicals (O_2^-), which were responsible for the observed MB degradation. In another case, Yu et al. [47] reported the efficient photocatalytic H_2 evolution activity of CQDs/P25 composite under both UV-vis and visible light ($\lambda > 450$ nm) irradiation without the aid of any noble metal cocatalyst. Different from pure P25, CQDs/P25 exhibited visible light-driven photocatalytic activity for H_2 production, ascribing to the photosensitizing ability of CQDs. Just like organic dyes, CQDs sensitized P25 into a visible light responsive "dyade" structure through the newly formed Ti—O—C bond, in

http://dx.doi.org/10.5772/intechopen.70801

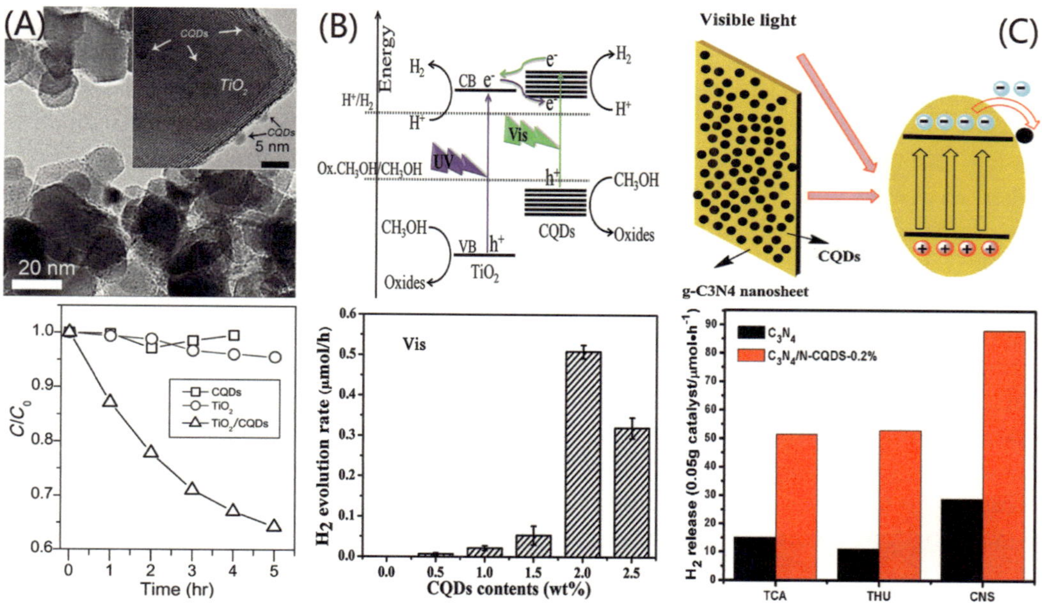

Figure 6. (A) TEM image of the TiO_2/CQDs hybrids and MB concentration (C/C_0) versus photocatalytic reaction time. (B) Schematic illustration for the H_2 production mechanism on the CQDs/TiO_2 under UV and visible light ($\lambda > 450$ nm), and photocatalytic H_2 evolution rates on CQDs/TiO_2 with varied CQDs contents under visible ($\lambda > 450$ nm) irradiation. (C) Schematic illustration of the C_3N_4/CQDs composites, and the phototcatalytic performance of C_3N_4 and 0.2 wt%-CQDs/C_3N_4 for the H_2 evolution rate under visible ($\lambda > 420$ nm) irradiation. Reprinted with permission [45, 47, 48]. Copyright 2015, Wiley-VCH; copyright 2014, Royal Society of Chemistry; copyright 2016, Elsevier.

which photo-induced electrons transferred from the excited CQDs to CB states of TiO_2, followed by reactions with protons for H_2 production (**Figure 6B**).

The characterized sp2 carbon composition of CQDs also contributes to the charge transfer ability as photosensitizers. In a recent report of Liu et al. [48], CQDs synthesized from bee pollens were used to couple with g-C_3N_4 nanosheets for a 2D/0D type photocatalysts via hydrothermal treatment. The composite has overcome the limited visible light absorption of C_3N_4, showing an optimal H_2 evolution of 88.1 μmol/h, 3.02 folds of pristine C_3N_4 nanosheets. Well dispersed CQDs uniformly anchored to the C_3N_4 nanosheets network via π-π stacking interactions were suggested to effectively expand the visible light absorption via photosensitization and hence suppress the recombination of photo-induced carriers (**Figure 6C**).

Overall, due to the excellent light harvesting ability over a broad spectrum, as well as the effective chemical binding toward semiconductor surface, carbon dots have shown their potential as alternatives to the scarce and toxic heavy-metal compounds that are currently used as photosensitizer for visible light harvesting.

4.3. Spectral converters

The UCPL of CQDs is of particular interests in this context, promising an emission of shorter wavelength photons upon the simultaneous absorption of two or multiple longer wavelength photons. Such a spectral converting ability of CQDs provides an attractive strategy for the design of visible light- or even NIR-responsive photocatalysts. A large number of recent work concerning about the photocatalytic applications of CQDs have suggested the contribution of their UCPL properties. Ke et al. [49] used TiO_2 photocatalysts modified with CQDs to obtain a 3.6 times higher rate of MB degradation under visible light irradiation ($\lambda > 420$ nm) when compared with pure TiO_2, and the enhanced photocatalysis was mainly ascribed to the increased amount of oxidative radicals in hypothesis of the up-conversion process from CQDs (**Figure 7A**). Similar results have also been observed with CQDs/g-C_3N_4 hybrids, as shown by Zhang et al. [50]. After impregnation of CQDs into g-C_3N_4, a 3.7-fold enhancement in phenol degradation rate constant can be obtained compared with that of the pristine g-C_3N_4 under visible light irradiation ($\lambda > 400$ nm). As suggested by the authors, light of $\lambda > 550$ nm could be converted to $\lambda < 460$ nm photons by CQDs, and subsequently excited g-C_3N_4 to generate electron-hole pairs (**Figure 7B**). UCPL character of CQDs that enhanced the production of excitons by extending the visible light-absorption region was supposed to be one important contribution to the observed enhanced photocatalysis, along with the efficient charge separation arising from the band structure alignment.

However, some researchers clarified that UCPL of CQDs might not be the main reason for the improved photocatalytic activity after CQDs introduction. In a recent work of Di et al. [51], improved photocatalytic degradation of ciprofloxacin on N-CQDs/BiPO4 composite

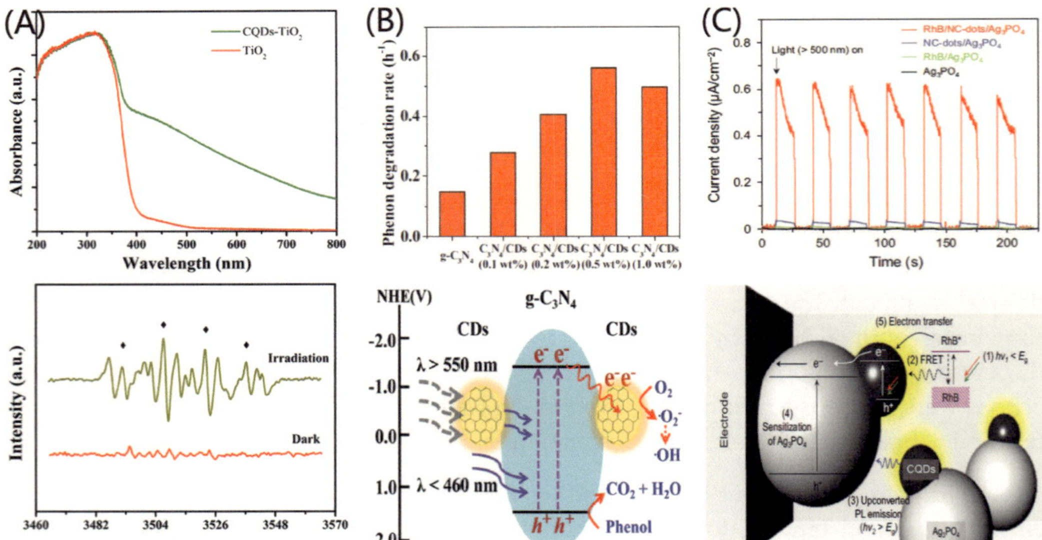

Figure 7. (A) UV-vis absorption spectra of TiO_2 and CQDs/TiO_2 hybrid, and the EPR spectra of hydroxyl radicals over the CQDs/TiO_2 hybrid under visible irradiation ($\lambda > 420$ nm). (B) Apparent rate constants of phenol degradation over CQDs modified C_3N_4 photocatalysts with varied loading amount, and the schematic illustration of the possible photocatalytic process on the C_3N_4/CQDs hybrid under visible light ($\lambda > 400$ nm). (C) Transient photocurrent responses at an applied potential of 0 V under visible irradiation ($\lambda > 500$ nm) for photocatalysts with different composition, and the schematic energy band diagram of the RhB/CQDs/Ag_3PO_4 nanocomposite. Reprinted with permission [49, 50, 52]. Copyright 2017, Elsevier; copyright 2016, Elsevier; copyright 2016, Elsevier.

was observed under UV irradiation, but only with negligible efficiency under visible light. Subsequently, the authors proposed that the up-converted PL may not play the crucial role for the improved activity in this system, instead, molecular oxygen activated by the oxygen defects on the surface of N-CQDs may be responsible for the organic removal under UV irradiation.

In fact, no direct evidence has been provided to confirm the contribution of UCPL to the improved photocatalytic activities postulated in many CQDs based photocatalytic systems. The generally poor efficiency of up-conversion process in most cases, if such an effect exists, may lead to its small influence compared to other optoelectronic contributions. Yu et al. [52] magnified the UCPL of CQDs by means of fluorescence resonance energy transfer (FRET)-assisted up-conversion process, in which rhodamine B (RhB) was combined with naked CQDs of 3.4 nm to make a FRET pair. The FRET effect was used to magnify the up-conversion of CQDs, with the up-converted photons subsequently sensitized Ag_3PO_4 particles and thus generating 18 times higher photocurrent at $\lambda > 500$ nm than those without RhB molecules (**Figure 7C**).

4.4. Sole photocatalyst

The origin of light absorption in CQDs stems from π-π* (C=C) and n-π* (C=O) transitions in the core and on the surface of the particles, respectively, which means that the optoelectronic properties of CQDs will be readily tuned by both size modulation and chemical modification. The size effect largely arises from the confinement, causing a separation of π and π* orbitals prominently when the domain size is less than 10 nm. At the same time, CQDs rich in surface oxygen groups are usually p-doped due to the larger electronegativity of oxygen atoms relative to carbon atoms. Replacing oxygen functional groups on the CQDs surface with nitrogen-containing groups can readily transform CQDs into an n-type semiconductor. In addition to surface modification, direct substitution with heteroatoms in the graphene lattice also greatly affects the optical and electronic properties of CQDs. These excellent properties have gained CDs with an adjustable band structure together with broadband light-absorption ability, which inspire their applications as efficient photocatalyst.

Although CQDs alone as photocatalysts show great potential, relatively few examples of CQDs solely with satisfying photocatalytic activity have been recorded to date, partially due to the relative poor electron transfer inside the CQDs. Doping with heteroatoms has been proved to be an effective way to promote electron transfer and thus the performance of the CQDs as photocatalyst. Cu–N co-doping of CQDs was achieved by Wu et al. [53] through an one-step pyrolytic synthesis with $Na_2[Cu(EDTA)]$ as precursor (**Figure 8A**). As a result of the Cu—N doping, the electron accepting and donating abilities of CQDs could be both enhanced, together with the increased electric conductivity. These merits ultimately facilitated the entire electron-transfer process in CQDs and further improved the photocatalytic oxidation of 1,4-dihydro-2,6-dimethylpyridine-3,5-dicarboxylate (1,4-DHP). In another case provided by Martindale et al. [54], N doped CQDs were synthesized by pyrolysis of aspartic acid (Asp) in air at 320°C. CQDs with N doping showed increased specific activity of 7950 μmol_{H2} $(g_{CD})^{-1}$ h^{-1} in the solar (AM1.5G) photocatalytic H_2 evolution system, with an

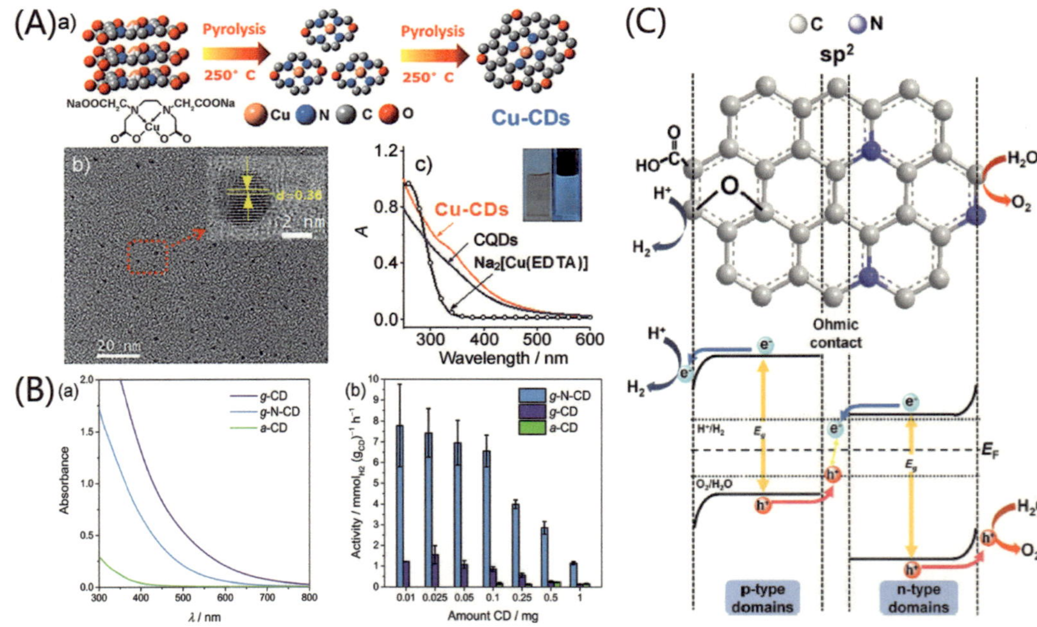

Figure 8. (A) (a) The synthetic procedure, (b) TEM images and (c) UV/vis absorption spectra of doped CQDs. (B) (a) UV/vis absorption spectra of N doped CQDs (g-N-CD), graphite CQDs (g-CD) and amorphous CQDs (a-CD), (b) photocatalytic H_2 generation with various amounts of CQDs (0.1–1 mg) under simulated solar irradiation (AM 1.5G, 100 mW cm^{-2}). Reprinted with permission [53–55]. Copyright 2015, Wiley-VCH; copyright 2017, Wiley-VCH; copyright 2014, Wiley-VCH.

order of magnitude improvement realized compared to undoped CQDs (**Figure 8B**). The authors suggested that nitrogen doping in CQDs increased the efficiency of hole-scavenging through electron donor effect and thereby significantly extended the lifetime of the photogenerated electrons.

Moreover, the electronic structure of CQDs can also be dramatically changed after heteroatomic doping. By treating GO in NH_3 at 500°C followed by oxidation with a modified Hummers' method, Yeh et al. [55] obtained N doped CQDs and applied as visible light ($\lambda > 420$ nm) driven photocatalyst for the overall water-splitting. Based on the electrochemical Mott-Schottky analysis, N doped CDs showed both p- and n-type conductivities, with the n-conductivity caused by embedding nitrogen atoms in the graphene frame and the p-conductivity by grafting oxygen functionalities on the graphene surface. The inherent sp2 clusters of CQDs served as the junction between the p- and n-domains to form buitin p-n diodes, which were responsible for an internal Z-scheme charge transfer at the QD interface, similar to that of biological photosynthesis (**Figure 8C**).

Therefore, by suitably adjusting the size, defect densities, atom doping and surface functional groups, it is possible to tune the electronic structure of CQDs and thus the optical response over a wide spectrum, allowing the potential application as effective photocatalysts with full utilization of the solar light.

5. Summary and outlook

After a decade of extensive investigation, CQDs have already been developed to be one of the most important members of modern nanocarbon family. Unexpected optoelectronic properties have been observed with this initially emergent nanocarbon material, including UV-visible absorbance, tunable PL and unique UCPL, together with electronic conductivity, showing the opportunity of further adjustment through surface functionalization and heteroatom doping. Specially, the excellent light harvesting capability and efficient photo-induced electron transfer ability make CQDs an exceptional candidate in visible light-driven photocatalytic applications, which play multifaceted roles as electron mediator, photosensitizer, spectral converter, or photocatalyst. Characterized by low cost, easy preparation, water soluble and nontoxic, CQDs have served as a common replacement for toxic heavy metal-based QDs and high-expense noble metals as co-catalysts. In addition, benefiting from the generally simple and economic synthesis of CQDs out of a wide spectrum of biomass carbon sources, the application of CQDs as photocatalytic component is of great potential in view of large-scale production especially for technological purpose.

Acknowledgements

This work was supported by the National Natural Science Foundation of China (51372266, 51572286, and 21503266) and the Outstanding Youth Fund of Jiangsu Province (BK20160011).

Author details

Shan Cong* and Zhigang Zhao

*Address all correspondence to: Scong2012@sinano.ac.cn

Key Lab of Nanodevices and Applications, Suzhou Institute of Nano-Tech and Nano-Bionics, Chinese Academy of Sciences (CAS), SuzhouPR, China

References

[1] Xu X, Ray R, Gu Y, Ploehn HJ, Gearheart L, Raker K, Scrivens WA. Electrophoretic analysis and purification of fluorescent single-walled carbon nanotube fragments. Journal of the American Chemical Society. 2004;**126**(40):12736-12737. DOI: 10.1021/ja040082h

[2] Zhou J, Booker C, Li R, Zhou X, Sham TK, Sun X, Ding Z. An electrochemical avenue to blue luminescent nanocrystals from multiwalled carbon nanotubes (MWCNTs). Journal of the American Chemical Society. 2007;**129**(4):744-745. DOI: 10.1021/ja0669070

[3] Zhang J, Yu SH. Carbon dots: Large-scale synthesis, sensing and bioimaging. Materials Today. 2016;**19**(7):382-393. DOI: 10.1016/j.mattod.2015.11.008

[4] Yuan F, Li S, Fan Z, Meng X, Fan L, Yang S. Shining carbon dots: Synthesis and biomedical and optoelectronic. Nano Today. 2016;**11**(5):565-586. DOI: 10.1016/j.nantod.2016.08.006

[5] Li H, Kang Z, Liu Y, Lee ST. Carbon nanodots: Synthesis, properties and applications. Journal of Materials Chemistry. 2012;**22**:24230-24253. DOI: 10.1039/C2JM34690G

[6] Lim SY, Shen W, Gao Z. Carbon quantum dots and their applications. Chemical Society Reviews. 2015;**44**:362-381. DOI: 10.1039/C4CS00269E

[7] Wang Y, Hu A. Carbon quantum dots: Synthesis, properties and applications. Journal of Materials Chemistry C. 2014;**2**:6921-6939. DOI: 10.1039/c4tc00988f

[8] Wang R, Lu KQ, Tang ZR, Xu YJ. Recent progress in carbon quantum dots: Synthesis, properties and applications in photocatalysis. Journal of Materials Chemistry A. 2015; **5**:3717-3734. DOI: 10.1039/c6ta08660h

[9] Zheng XT, Ananthanarayanan A, Luo KQ, Chen P. Glowing graphene quantum dots and carbon dots: Properties, syntheses, and biological applications. Small. 2015;**11**(14):1620-1636. DOI: 10.1002/smll.201402648

[10] Lu J, Yang J-X, Wang J, Lim A, Wang S, Loh KP. One-pot synthesis of fluorescent carbon nanoribbons, nanoparticles, and graphene by the exfoliation of graphite in ionic liquids. ACS Nano. 2009;**3**(8):2367-2375. DOI: 10.1021/nn900546b

[11] Pan D, Zhang J, Li Z, Wu M. Hydrothermal route for Cutting graphene sheets into blue-luminescent graphene quantum dots. Advanced Materials. 2009;**22**(6):734-738. DOI: 10.1002/adma.200902825

[12] Tao H, Yang K, Ma Z, Wan J, Zhang Y, Kang Z, Liu Z. In vivo NIR fluorescence imaging, biodistribution, and toxicology of photoluminescent carbon dots produced from carbon nanotubes and graphite. Small. 2012;**8**(2):281-290. DOI: 10.1002/smll.201101706

[13] Li Y, Hu Y, Zhao Y, Shi G, Deng L, Hou Y, Qu L. An electrochemical avenue to green-luminescent graphene quantum dots as potential electron-Acceptors for photovoltaics. Advanced Materials. 2011;**23**(6):776-780. DOI: 10.1002/adma.201003819

[14] Shinde DB, Pillai VK. Electrochemical preparation of luminescent graphene quantum dots from multiwalled carbon nanotubes. Chemistry—A European Journal. 2012;**18**(39):12522-12528. DOI: 10.1002/chem.201201043

[15] Lu Q, Wu C, Liu D, Wang H, Su W, Li H, Zhang Y, Yao S. A facile and simple method for synthesis of graphene oxide quantum dots from black carbon. Green Chemistry. 2017; **19**:900-904. DOI: 10.1039/C6GC03092K

[16] Qiao Z-A, Wang Y, Gao Y, Li H, Dai T, Liu Y, Huo Q. Commercially activated carbon as the source for producing multicolor photoluminescent carbon dots by chemical oxidation. Chemical Communications. 2010;**46**:8812-8814. DOI: 10.1039/C0CC02724C

[17] Yu H, Li X, Zeng X, Lu Y. Preparation of carbon dots by non-focusing pulsed laser irradiation in toluene. Chemical Communications. 2016;**52**:819-822. DOI: 10.1039/C5CC08384B

[18] Li JY, Liu Y, Shu QW, Liang JM, Zhang F, Chen XP, Deng XY, Swihart MT, Tan KJ. One-pot hydrothermal synthesis of carbon dots with efficient up- and down-converted photoluminescence for the sensitive detection of morin in a dual-readout assay. Langmuir. 2017;**33**(4):1043-1050. DOI: 10.1021/acs.langmuir.6b04225

[19] Deng J, Lu Q, Mi N, Li H, Liu M, Xu M, Tan L, Xie Q, Zhang Y, Yao S. Electrochemical synthesis of carbon Nanodots directly from alcohols. Chemistry—A European Journal. 2014;**20**(17):4993-4999. DOI: 10.1002/chem.201304869

[20] Sun S, Zhang L, Jiang K, Aiguo W, Lin H. Toward high-efficient red emissive carbon dots: Facile preparation, unique properties, and applications as multifunctional theranostic agents. Chemistry of Materials. 2016;**28**(23):8659-8668. DOI: 10.1021/acs.chemmater.6b03695

[21] Chen X, Zhang W, Wang Q, Fan J. C_8-structured carbon quantum dots: Synthesis, blue and green double luminescence, and origins of surface defects. Carbon. 2014;**79**:165-173. DOI: 10.1016/j.carbon.2014.07.056

[22] Yang Z-C, Wang M, Yong AM, Wong SY, Zhang X-H, Tan H, Chang AY, Li X, Wang J. Intrinsically fluorescent carbon dots with tunable emission derived from hydrothermal treatment of glucose in the presence of monopotassium phosphate. Chemical Communications. 2011;**47**:11615-11617. DOI: 10.1039/C1CC14860E

[23] da Silva Souza DR, de Mesquita JP, Lago RM, Caminhas LD, Pereira FV. Cellulose nanocrystals: A versatile precursor for the preparation of different carbon structures and luminescent carbon dots. Industrial Crops and Products. 2016;**93**:121-128. DOI: 10.1016/j.indcrop.2016.04.073

[24] Choi Y, Thongsai N, Chae A, Jo S, Kang EB, Paoprasert P, Park SY, In I. Microwave-assisted synthesis of luminescent and biocompatible lysine-based carbon quantum dots. Journal of Industrial and Engineering Chemistry. 2017;**47**:329-335. DOI: 10.1016/j.jiec.2016.12.002

[25] Sahu S, Behera B, Maiti TK, Mohapatra S. Simple one-step synthesis of highly luminescent carbon dots from orange juice: Application as excellent bio-imaging agents. Chemical Communications. 2012;**48**:8835-8837. DOI: 10.1039/C2CC33796G

[26] Wang J, Wang C-F, Chen S. Amphiphilic egg-derived carbon dots: Rapid plasma fabrication, pyrolysis process, and multicolor printing patterns. Angewandte Chemie. 2012;**124**(37):9431-9435. DOI: 10.1002/ange.201204381

[27] Guo Y, Zhang L, Cao F, Leng Y. Thermal treatment of hair for the synthesis of sustainable carbon quantum dots and the applications for sensing Hg^{2+}. Scientific Reports. 2016;**6**:35795. DOI: 10.1038/srep35795

[28] Bao L, Zhang Z-L, Tian Z-Q, Li Z, Liu C, Yi L, Qi B, Pang D-W. Electrochemical tuning of luminescent carbon nanodots: From preparation to luminescence mechanism. Advanced Materials. 2011;**23**(48):5801-5806. DOI: 10.1002/adma.201102866

[29] Tan X, Li Y, Li X, Zhou S, Fan L, Yang S. Electrochemical synthesis of small-sized red fluorescent graphene quantum dots as a bioimaging platform. Chemical Communications. 2015;**51**:2544-2546. DOI: 10.1039/C4CC09332A

[30] Li H, He X, Kang Z, Huang H, Liu Y, Liu J, Lian S, Tsang CHA, Yang X, Lee S-T. Water-soluble fluorescent carbon quantum dots and photocatalyst design. Angewandte Chemie, International Edition. 2010;**49**(26):4430-4434. DOI: 10.1002/anie.200906154

[31] Li Y, Shu H, Niu X, Wang J. Electronic and optical properties of edge-functionalized graphene quantum dots and the underlying mechanism. Journal of Physical Chemistry C. 2015;**119**(44):24950-24957. DOI: 10.1021/acs.jpcc.5b05935

[32] Li X, Lau SP, Tang L, Ji R, Yang P. Sulphur doping: A facile approach to tune the electronic structure and optical properties of graphene quantum dots. Nanoscale. 2014;**6**:5323-5328. DOI: 10.1039/C4NR00693C

[33] Lin L, Zhang S. Creating high yield water soluble luminescent graphene quantum dots via exfoliating and disintegrating carbon nanotubes and graphite flakes. Chemical Communications. 2012;**48**:10177-10179. DOI: 10.1039/C2CC35559K

[34] Peng H, Travas-Sejdic J. Simple aqueous solution route to luminescent carbogenic dots from carbohydrates. Chemistry of Materials. 2009;**21**(23):5563-5565. DOI: 10.1021/cm901593y

[35] Dan Q, Zheng M, Peng D, Zhou Y, Zhang L, Di L, Tan H, Zhao Z, Xie Z, Sun Z. Highly luminescent S, N co-doped graphene quantum dots with broad visible absorption bands for visible light photocatalysts. Nanoscale. 2013;**5**:12272-12277. DOI: 10.1039/C3NR04402E

[36] Liu R, Dongqing W, Liu S, Koynov K, Knoll W, Li Q. An aqueous route to multicolor photoluminescent carbon dots using silica spheres as carriers. Angewandte Chemie. 2009;**121**(25):4668-4671. DOI: 10.1002/ange.200900652

[37] Li C, Meziani MJ, Sahu S, Sun Y-P. Photoluminescence properties of graphene versus other carbon nanomaterials. Accounts of Chemical Research. 2013;**46**(1):171-180. DOI: 10.1021/ar300128j

[38] Li C, Wang X, Meziani MJ, Fushen L, Wang H, Luo PG, Yi L, Harruff BA, Monica Veca L, Murray D, Xie S-Y, Sun Y-P. Carbon dots for multiphoton bioimaging. Journal of the American Chemical Society. 2007;**129**(37):11318-11319. DOI: 10.1021/ja073527l

[39] Gan Z, Xinglong W, Zhou G, Shen J, Chu PK. Is there real upconversion photoluminescence from graphene quantum dots? Advanced Optical Materials. 2013;**1**(8):554-558. DOI: 10.1002/adom.201300152

[40] Shen J, Zhu Y, Chen C, Yang X, Li C. Facile preparation and upconversion luminescence of graphene quantum dots. Chemical Communications. 2011;**47**:2580-2582. DOI: 10.1039/C0CC04812G

[41] Wen X, Pyng Y, Toh Y-R, Ma X, Tang J. On the upconversion fluorescence in carbon nanodots and graphene quantum dots. Chemical Communications. 2014;**50**:4703-4706. DOI: 10.1039/c4cc01213e

[42] Miao X, Ji Z, Wu J, Shen X, Wang J, Kong L, Liu M, Song C. g-C3N4/AgBr nanocomposite decorated with carbon dots as a highly efficient visible-light-driven photocatalyst. Journal of Colloid and Interface Science. 2017;**502**:24-32. DOI: 10.1016/j.jcis.2017.04.087

[43] Huang Y, Liang Y, Rao Y, Zhu D, Cao J-j, Shen Z, Ho W, Lee SC. Environment-friendly carbon quantum dots/$ZnFe_2O_4$ photocatalysts: Characterization, biocompatibility, and mechanisms for NO removal. Environmental Science & Technology. 2017;**51**(5):2924-2933. DOI: 10.1021/acs.est.6b04460

[44] Xiuqin W, Zhao J, Guo S, Wang L, Shi W, Huang H, Liu Y, Kang Z. Carbon dot and $BiVO_4$ quantum dot composites for overall water splitting via a two-electron pathway. Nanoscale. 2016;**8**:17314-17321. DOI: 10.1039/C6NR05864G

[45] Zhang J, Abbasi F, Claverie J. An efficient templating approach for the synthesis of redispersible size-controllable carbon quantum dots from graphitic polymeric micelles. Chemistry—A European Journal. 2015;**21**(43):15142-15147. DOI: 10.1002/chem.201502158

[46] Qiong W, Li W, Yanjiao W, Huang Z, Liu S. Pentosan-derived water-soluble carbon nano dots with substantial fluorescence: Properties and application as a photosensitizer. Applied Surface Science. 2014;**315**:66-72. DOI: 10.1016/j.apsusc.2014.06.127

[47] Huijun Y, Zhao Y, Zhou C, Shang L, Peng Y, Cao Y, Li-Zhu W, Tunga C-H, Zhang T. Carbon quantum dots/TiO_2 composites for efficient photocatalytic hydrogen evolution. Journal of Materials Chemistry A. 2014;**2**:3344-3351. DOI: 10.1039/C3TA14108J

[48] Liu Q, Chen T, Guo Y, Zhang Z, Fang X. Ultrathin g-C_3N_4 nanosheets coupled with carbon nanodots as 2D/0D composites for efficient photocatalytic H_2 evolution. Applied Catalysis B: Environmental. 2016;**193**:248-258. DOI: 10.1016/j.apcatb.2016.04.034

[49] Ke J, Li X, Zhao Q, Liu B, Liu S, Wang S. Upconversion carbon quantum dots as visible light responsive component for efficient enhancement of photocatalytic performance. Journal of Colloid and Interface Science. 2017;**496**:425-433. DOI: 10.1016/j.jcis.2017.01.121

[50] Zhang H, Zhao L, Geng F, Guo L-H, Wan B, Yang Y. Carbon dots decorated graphitic carbon nitride as an efficient metal-free photocatalyst for phenol degradation. Applied Catalysis B: Environmental. 2016;**180**:656-662. DOI: 10.1016/j.apcatb.2015.06.056

[51] Di J, Xia J, Chen X, Ji M, Yin S, Qi Z, Li H. Tunable oxygen activation induced by oxygen defects in nitrogen doped carbon quantum dots for sustainable boosting photocatalysis. Carbon. 2017;**114**:601-607. DOI: 10.1016/j.carbon.2016.12.030

[52] Yu S, Lee SY, Umh HN, Yi J. Energy conversion of sub-band-gap light using naked carbon nanodots and rhodamine B. Nano Energy. 2016;**26**:479-487. DOI: 10.1016/j.nanoen.2016.06.008

[53] Wu W, Zhan L, Fan W, Song J, Li X, Li Z, Wang R, Zhang J, Zheng J, Wu M, Zeng H. Cu–N dopants boost electron transfer and photooxidation reactions of carbon dots. Angewandte Chemie, International Edition. 2015;**54**(22):6540-6544. DOI: 10.1002/anie.201501912

[54] Martindale BCM, Hutton GAM, Caputo CA, Prantl S, Godin R, Durrant JR, Reisner E. Enhancing light absorption and charge transfer efficiency in carbon dots through graphitization and core nitrogen doping. Angewandte Chemie, International Edition. 2017;**56**(23):6459-6463. DOI: 10.1002/anie.201700949

[55] Yeh TF, Teng CY, Chen SJ, Teng H. Nitrogen-doped graphene oxide quantum dots as photocatalysts for overall water-splitting under visible light illumination. Advanced Materials. 2014;**26**(20):3297-3303. DOI: 10.1002/adma.201305299

Chapter 5

Graphene-Semiconductor Composites as Visible Light-Induced Photocatalyst

Ozlem Altintas Yildirim and Teoman Ozturk

Additional information is available at the end of the chapter

http://dx.doi.org/10.5772/intechopen.70709

Abstract

Graphene-based composites produced by the incorporation of graphene into suitable semiconductors doped with various metals enable to induce the unique properties of the graphene, such as extended light absorption range, charge separation, and so high dye adsorption capacity. Therefore, graphene-based composites can provide to enhance the overall photocatalytic performance of the base semiconductor under the visible-light, and to open up new pathways to high-performance photocatalyst for the future applications. This chapter of the book focuses on the structural and optical properties of the graphene-semiconductor-based composite structure. Furthermore, final photocatalytic properties of the graphene-based composites can be controlled by applying different synthesis routes. Common synthesis methods of the graphene-semiconductor composites such as sol-gel, solution mixing, *in situ* growth, hydrothermal growth, and solvothermal method are discussed on the resultant visible-light photocatalytic property of the these composites. At the same time, doping of the graphene-semiconductor material with metal ions also allows an improvement of the visible light-induced photocatalytic activity. Therefore, studies related with the effect of the dopant agent on the visible light photocatalytic activity are also reviewed in this chapter.

Keywords: graphene-semiconductor composites, photocatalyst, visible light degradation, structural-electronic properties, synthesis methods

1. Introduction

Environmental water contaminations originated from the organic dye compounds used in textile and paint industries and household chemicals are a continuous threat to human health. Generally, water treatment methods include primary and secondary treatment steps. While primary treatment of water is designed to physically remove organic solids by sedimentation and flotation, secondary treatment is aimed to biologically remove the residual organics from wastewater by utilizing microorganisms in a controlled environment [1]. As an advanced treatment step, desalination technologies such as reverse osmosis, electrodialysis, ion exchange, and freeze desalination were applied. However, the above-listed traditional biological treatment techniques can be ineffective in decolorizing dye molecules. Due to the specific needs during the degradation of organic pollutants, there has been a strong interest in the development of efficient treatment method.

Recently, in order to degrade organic dye compounds, many researchers have studied on the photocatalyst material which is usually a wide band gap semiconductor in the presence of solar energy. The photocatalytic reaction mechanism is based on the ability to create electron-hole pairs under photo illumination. For photocatalytic devices, sunlight, which is abundantly available as natural irradiation energy, and UV irradiation sources can be conveniently exploited for the irradiation of semiconducting materials. For practical applications, development of an efficient, reproducible, and cost-effective visible light-induced photocatalyst is desirable for the large-scale production of catalyst. Recently, many research papers are dedicated to the semiconductor-based [zinc oxide (ZnO), cadmium sulfide (CdS), cadmium selenide (CdSe), iron oxide (Fe_2O_3), tin oxide (SnO_2), zinc sulfide (ZnS), etc.] [2], carbonaceous-based (nanotubes, fullerenes, graphene, etc.) [3], and semiconductor-carbonaceous composite photocatalysts [4]. Over the recent decades, semiconductor-based photocatalysts have attracted great attention in water purification because of their high photosensitivity, environment-friendly non-toxic nature, and low cost [5].

Although semiconductor-based photocatalysts, especially TiO_2, have been recognized as the most effective catalysts and are widely used in the removal of toxic or hazardous organic pollutants, several fundamental issues must be addressed before the photocatalysts are efficiently used for the industrial applications. For example, large bandgap of the TiO_2 photocatalyst provides only absorbance of UV light with a wavelength of less than 385 nm, which accounts for less than 10% of solar irradiation. In addition, fast recombination of electron-hole pairs limits the applicability of TiO_2 as photocatalyst. These problems are the cause of low photosensitivity and high cost, and makes these types of photocatalyst unfavorable for large scale production.

In the family of carbon nanomaterials, graphene-based photocatalyst activated under the visible light seems to be the preferred candidates because of their outstanding advantages such as low cost, innocuous nature, ease of availability/processing, and suitability of their structure to create composites with semiconductors [6]. Graphene is a two-dimensional material having sp^2-bonded carbon atoms tightly packed into a two-dimensional honeycomb structure. After discovering by Novoselov et al. in 2004 [7], graphene has been attracting more and more attention because of their unique properties such as excellent mobility of charge carriers at room temperature (100,000 $cm^2 V^{-1} s^{-1}$) [8], theoretically large surface area (2630 $m^2 g^{-1}$) [9], optical transparency [10], excellent mechanical strength (2.4 ± 0.4 TPa) [11], high thermal conductivity (~5000 $W m^{-1} K^{-1}$) [12], exhibiting half-integer room-temperature quantum Hall effect [13], capacity of sustaining large electrical current density (10^8 A cm^{-2}) [14], etc. It has attracted a great attention for their potential applications in variety fields such as catalyst

[15], sensors [16], nano electronics [17], energy conversion [18], and water treatment [4]. In the past few decades, there is an ever-increasing interest in the photocatalytic applications of graphene [19].

To improve the photocatalytic efficiency of the semiconductor-based metal oxides and graphene-based photocatalysts under the visible-light, integration of semiconductor metal oxides with graphene has been one the most popular approach due to theoretically large specific-surface area and high intrinsic electron mobility of graphene at room temperature [3]. Graphene behaves as an electron acceptor in this type composite photocatalysts. Generally, degradation of the organic dye molecules takes place with anchoring onto the surface of semiconducting particles by means of covalent bonding and ion-pair type association. In the literature, different types of graphene-based composites (graphene-semiconductor, graphene oxide (GO)-semiconductor, and reduced graphene oxide-semiconductor; as a semiconductor material TiO_2, ZnO, CdS, Zn_2SnO_4, etc.) can be found. Graphene-semiconductor composites have been fabricated through simple mixing and/or sonication, sol-gel process, liquid-phase, hydrothermal, and solvothermal methods. Furthermore, as a strategy to improve visible-light photocatalytic efficiency of graphene-based composites, doping semiconductor with metal ions is an effective way. In this chapter of the book, structural and optical properties of the graphene-semiconductor composites, general synthesis methods, and photocatalytic properties of these composites under the visible light are discussed. Finally, the effect of the doping on the resultant visible light-induced photocatalytic activity of graphene-semiconductor composites is reported.

2. The graphene-semiconductor composites; structural and optical properties

Over the last decades, spreading of diseases due to organic dye compounds caused by noxious chemicals has become a global public health issue. Thus, scientists have focused on the studies of the wastewater treatment and antibacterial applications required to use a photocatalyst material. In the studies to prevent the spreading of the dye molecules, carbonaceous materials draw attention because of offering high photocatalytic activity [6].

Carbon is one of the most abundant elements in the world and is necessary for life in the nature. Carbon has different forms or allotropes which include buckminsterfullerene (or just simply fullerene), diamond, graphite, etc. The discovery of buckminsterfullerene (C60) [20] which is a cluster of 60 carbon atoms brought the Nobel Prize in Chemistry in 1996 to Kroto, Curl, and Smalley. On the other hand, diamond is considered as the hardest material, whereas graphite is extremely soft. Graphite is a naturally occurring mineral and has parallel carbon atom layers which are named as graphene, so that graphene can be defined as a single and flat layer of graphite which consists of sp^2-bonded carbon atoms packed into a two-dimensional (2D) honeycomb lattice. Graphene was first isolated as a single-layer from graphite in 2004 by Novoselov et al. using the mechanical exfoliation technique [7]. This technique is based on the repeated peeling of thin flakes from a piece of graphite with a scotch tape. This success has led to an incredible increase in theoretical and experimental studies about graphene in the recent years. Geim and Novoselov were awarded the Nobel Prize in Physics 2010 because of the producing, isolating, identifying and characterizing graphene.

Apart from the exfoliation method, alternative ways for the producing of graphene have been developed: epitaxial growth on silicon carbide is based on the evaporating silicon from SiC with heating and leaving a thin carbon layer [21]. Chemical vapor deposition on metal surfaces is one of the fabricate methods of graphene. A transition metal (nickel) film is exposed to a hydrocarbon gas at high temperature under ambient pressure. Cooling the sample, solubility of carbon in the transition metal film decreases and an ultrathin graphene layer is formed over the metal surface [22, 23].

Besides these methods, the chemical reduction of graphene oxide (GO) derived from chemical exfoliation of graphite is one of efficient and low cost ways in order to obtain large scale graphene which was first reported by Ruoff's group [24–27]. GO is a strongly oxidized form of graphene which can be obtained using oxidizing agents from graphite oxide. Graphite oxide is generally obtained with Hummers' method [28] which involves the oxidation of graphite powder in the mixture of concentrated sulfuric acid (H_2SO_4), sodium nitrate ($NaNO_3$), and potassium permanganate ($KMnO_4$). After oxidation, the layers of graphite oxide, which consists of GO sheets exhibit hydrophilicity, allows water to intercalate between the sheets and disperse them. So, the exfoliation of graphite oxide is facilitated and a layer graphene or GO is produced via ultrasonic treatment in aqueous media. GO contains various types of oxygen containing functional groups such as hydroxyl, epoxide, and carbonyl groups caused by heavily oxidation. Due to the disruption of the conjugated structure by functionalization, covalent bonds of carbon and oxygen atoms become more intense and the resistance of the layer increases, so the conductivity of GO decreases. For this reason, a more efficient chemical reduction method is required in order to obtain GO with high conductivity. In other words, GO needs to be reduced by removing the oxygen-bearing group. The term of the reduced GO (rGO) is used to refer to as chemically modified graphene, chemically derived graphene, or reduced graphene [29].

There are various GO reduction methods which can be achieved through chemical, thermal, or electrochemical ways [30, 31]. In order to obtain rGO with chemical way, reducing agents are used such as hydrazine (N_2H_4) [27], hydroquinone (HQ) [32] sodium borohydride ($NaBH_4$) [33], and hydriodic acid (HI) [34]. Hydrazine, which is one of the most widely used reducing agents, has been used to reduce GO in order to remove the functional groups. However, hydrazine is highly toxic and incorporates C—N groups which increase the resistance of GO sheets due to scattering electrons. During the thermal reduction, GO is exposed to a heating process with the evolved gases. Decomposing of the epoxy and hydroxyl groups with temperature generates high pressures that separate the graphene sheets [35–37]. However, thermal reduction damages the structure of graphene sheets and causes mass loss [31, 35].

Another method is electrochemical reduction, where the electron behaves as the reducing agent. In the electrochemical approach, a thin GO film is deposited onto the surface of the substrates such as flexible plastic, glass, indium tin oxide, and glassy carbon [38, 39]. Reference and auxiliary electrodes are placed and the electrochemical reduction process is performed on the GO with cyclic voltammetry in a buffer electrolyte. Oxygen-containing functional groups in GO can be reduced by the applying various voltages.

Exhibiting of the photocatalytic properties of the graphene-based materials depends on the creating free radicals in the photocatalytic reaction. However, fast recombination of the electron-hole pair, which is formed under the light source, is the major disadvantage of the using of the graphene as a photocatalyst material. In order to overcome this problem, chemical modification of the graphene is done with bioactive nanoparticles or light sensitive materials. Due to the expensive cost of gold and silver, which are commonly used as nanoparticles [40, 41], the applications of the photocatalyst composite formed with these elements and graphene is limited in large scale applications. The chemical modification of the graphene with light sensitive materials is another way to improve photocatalytic activity of graphene. TiO_2 is the light sensitive material used most commonly up till now. However, an alternative material which will form a composite structure with graphene is still being sought due to small portion solar light usage of TiO_2.

Since the seminal discovery of water splitting on TiO_2 with photoelectrolysis by Fujishima and Honda in 1972 [42], there is a growing interest in the photocatalysis of semiconductors. Many types of semiconductor photocatalysts have been reported up to now including TiO_2, ZnO, CdS, CdSe, Fe_2O_3, SnO_2, ZnS, etc. Among the many different semiconductor materials, TiO_2 has been widely used in the light-induced photochemical reactions because of its low cost, nontoxicity, chemical stability, and its highly oxidizing. However, the applications of TiO_2 are restricted due to fast recombination rate of photogenerated electron-hole pairs and its large band gap (3.2 eV) that requires the UV light irradiation. **Figure 1** shows a schematic illustration of the photocatalytic reaction mechanism. With the absorption of the photon energy, which has greater than that of the band gap of the semiconductor material, electron in the valance band becomes excited and jumps to the conduction band. Hence, a positively charged hole forms in the valence band. As a result of the recombination with the electron, this the positively charged hole can oxidize water molecules and provide to form hyper-reactive hydroxyl free radicals (•OH). These radicals induce the photodegradation of the dye molecules.

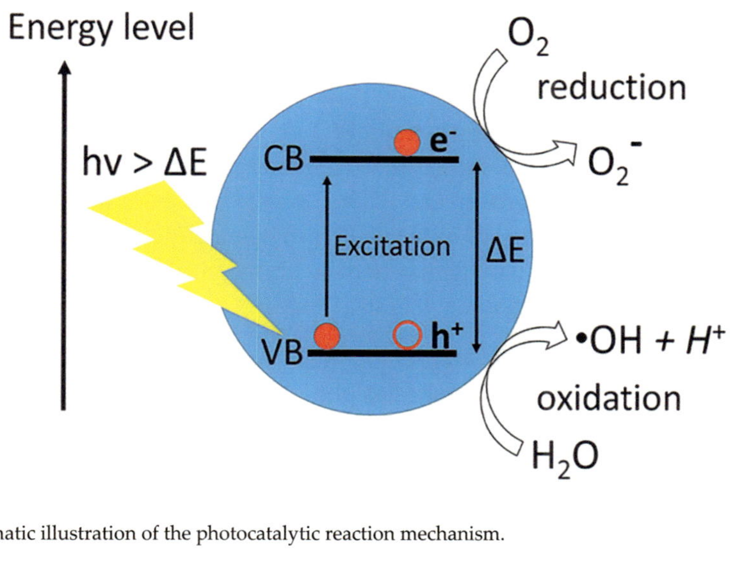

Figure 1. A schematic illustration of the photocatalytic reaction mechanism.

There has been considerable interest on the enhancement of photocatalytic activity by the combination of carbonaceous materials with semiconductors because of its extraordinary structural and electronic properties. The pioneering work of Williams et al. about UV-induced photocatalytic reduction of GO [43] has increased the interest in graphene-based semiconductor photocatalysts. They exposed a mixture of graphene oxide and TiO_2 to UV light in ethanol. When the holes that occur as a result of excitation are consumed to produce ethoxy radicals, the electrons accumulate within TiO_2 particles and reduce the oxygen containing functional groups on GO surface. Zhang et al. observed a significant enhancement in the photocatalytic activity of the photodegradation of methylene blue (MB) with chemically bonded graphene TiO_2 P25 (with particle size around 25 nm) nanocomposite photocatalyst [44]. These nanocomposite photocatalysts were obtained with the reduction of GO and loading of P25 using hydrothermal method.

Photocatalytic enhancement in the graphene-semiconductor composites can be explained with three mechanisms due to unique properties of graphene: (i) extension the wavelength of the absorbed light due to the chemical bonds which narrow the band gap of semiconductor and transparency of graphene; (ii) suppression of the recombination of photogenerated electron hole pairs. The excited electrons transfer from the conduction band of semiconductor to graphene which is an acceptor due to its long π conjugation structure [45]. With fast charge transportation and effective separation of electron-hole pairs, the charge recombination is suppressed; (iii) increasing absorptivity of pollutants. Model dyes or contaminant molecules usually have an aromatic structure and they could be conjugated to the aromatic regions of the graphene. This absorptivity increases the amount of contaminant molecules on the catalyst and improves a significant increase of photocatalytic degradation [44–46]. The role of adsorption of graphene in photocatalysis was studied by Liu et al. [47]. They have shown that rGO captures dye molecules from water through adsorption and photoinduced electrons from TiO_2 through surface during photocatalytic process. The photocatalytic mechanism in graphene-TiO_2 composites can be briefly summarized as follows:

$$TiO_2 + h\nu \rightarrow TiO_2\,(h^+ + e^-)$$

$$TiO_2\,(e^-) + O_2 + \text{graphene} \rightarrow TiO_2 + O_2^{\cdot-} + \text{graphene}\,(e^-)$$

$$TiO_2\,(h^+) + OH^- \rightarrow TiO_2 + \bullet OH$$

In order to increase the photocatalytic activity of TiO_2, it is required that the slowing of the charge recombination rate and the reduction of the energy band gap. These processes have been done with increasing the porosity and surface area [48], doping with metals [49] or non-metals [50], and surface coupling with metals [51] or semiconductors [52]. While there is a great challenge and effort to explain the mechanism of photocatalytic enhancement observed in graphene-semiconductor composites, it is still unclear.

Apart from TiO_2, there are some semiconductors which have been studied in the photocatalysis of graphene-semiconductors composites as photocatalysts. One of these semiconductors, ZnO, is a promising candidate for environmental contaminations due to its strong oxidizing power, good photocatalytic activity, and low cost. The crystal structures of ZnO exist in rock-salt, zinc blende, and wurtzite (hexagonal) forms. Wurtzite is the most commonly used and thermodynamically stable phase of ZnO [53]. ZnO has similar characteristics to TiO_2 due to its large band gap (3.37 eV) and UV light sensitivity. ZnO has been used in the photocatalytic reduction of GO in ethanol by Williams and Kamat [54]. Similar to mechanism of graphene-TiO_2 composites, electron-hole pairs generated by UV excited ZnO create electron-hole pairs due to UV irradiation as a photocatalyst and degrade the pollutants [55].

Although there have been a lot progress in the photocatalysis of graphene-based semiconductor composites, very small portion of the solar spectrum is utilized due to UV light sensitive semiconductors. In recent years, semiconductors such as SnO_2 [56], $BiVO_4$ [57], CdS [58], and CdSe [59] have been used as photocatalysts in graphene-semiconductor composites under visible light. These semiconductors generally have been chosen due to their proportional narrow band gaps. For example, the energy band gap of CdSe is 1.67 eV, so it can be used under visible light in order to use solar spectrum efficiently. Also, use of graphene enhances the photoactivity of graphene-semiconductor composites due to separation of charges, suppression of the charge recombination, and facilitation of charge transfer. There is a great challenge to improve the efficient conversion of solar spectrum in the visible light region photocatalysis which will be mentioned in section 4.

3. Common synthesis methods of the graphene-semiconductor composites

At the present time, many different methods have been developed to synthesize graphene-semiconductor composites. Most commonly used synthesis methods are sol-gel, solution mixing, *in situ* growth, hydrothermal, and solvothermal methods. More detailed explanations of these syntheses are given below.

3.1. Sol-gel method

Sol-gel method is the process of the suspending of particles in a liquid and placing of these particles onto a surface from a solution. Zhang et al. [60] synthesized graphene-TiO_2 composites employing a sol-gel method using tetrabutyl titanate (TBOT) and GO as starting materials. At first, they dispersed the graphene sheets into ethanol solvent, and then dropped TBOT into the suspension. The obtained sol was dried to form precursor. Finally, graphene/TiO_2 composites were synthesized by annealing the precursor for the crystallization. They investigated the photocatalytic activity of the graphene-TiO_2 composites using hydrogen evolution from water photo-splitting under UV-Vis illumination. Liu et al. [61] reported a preparation of graphene-TiO_2 composites using titanium precursors and rGO by a sol-gel method and observed an enhancement in the photocatalytic activity under visible light.

3.2. Solution mixing method

Solution mixing method is one of the most common and easiest methods for the preparation of graphene/semiconductor composites which are prepared using the suspensions of graphene and semiconductor nanoparticles. Paek et al. [62] prepared graphene-SnO_2 composites using solution mixing method. They first prepared graphene sheets via the chemical reduction of graphite oxide, and then obtained SnO_2 nanoparticles through the controlled hydrolysis of $SnCl_4$ with NaOH. Reduced graphene sheets were dispersed in the ethylene glycol, and reassembled in the presence of SnO_2 nanoparticles to form the composites. Bell et al. [63] prepared rGO-TiO_2 composites by mixing suspensions of TiO_2 powder with the GO suspension ultrasonically and exposing the suspension to UV light. Sonication has been used to dissolve GO in non-polar solvents. Using a similar technique, Akhavan and Ghaderi [64] prepared the graphene-TiO_2 thin films and studied the antibacterial activity of the graphene-TiO_2 composite thin films under solar light irradiation.

3.3. *In situ* growth method

In situ is derived from Latin meaning in the reaction mixture in chemistry. *In situ* growth is based on the direct reaction of graphene and semiconductor nanoparticles in solution, and widely used to prepare graphene-semiconductor composites. Lambert et al. [65] synthesized GO-TiO_2 composites via the hydrolysis of TiF_4 in the presence of aqueous dispersions of GO then obtained rGO-TiO_2 adding hydrazine after sonication. Zhang et al. [66] prepared rGO-SnO_2 and rGO-TiO_2 composites using *in situ* method. They reduced GO dispersion to rGO with adding $SnCl_2$ or $TiCl_3$. Sn^{2+} and Ti^{3+} ions were converted to SnO_2 or TiO_2 nanoparticles and rGO-semiconductor composites were studied for the photocatalytic properties for degrading rhodamine B (RhB) under visible light. Li and Cao [67] obtained graphene-ZnO composites via a chemical deposition route. GO suspension was mixed with Zn^{2+} ions containing solution then NaOH was added to the mixture. The powder obtained from the separated and dried solution was dispersed in $NaBH_4$ solution. After hydrothermal treatment, graphene-ZnO composites were obtained and used in the degradation of RhB under UV or visible light irradiation.

3.4. Hydrothermal growth method

Hydrothermal method can be defined as the process of growing crystals in the presence of water as a solvent at high temperature and pressure in an autoclave which is a closed reaction vessel. This method is one of the conventional methods in order to obtain graphene-semiconductor composites. Zhang et al. [44] obtained a chemically bonded graphene-TiO_2 (P25) nanocomposite with graphene oxide and P25, using a facile one-step hydrothermal method. In their work, GO was dissolved in a solution of distilled water and P25 was added to the solution. Then suspension was placed in an autoclave so GO was reduced and P25 nanoparticles were deposited on the graphene simultaneously. They measured an enhancement in the reaction rate of the photodegradation of MB using graphene-P25 nanocomposites under UV and visible light. Liu et al. [68] demonstrated a one-pot hydrothermal approach to prepare rGO-ZnO composites and applied these composites as photocatalyst in order to decompose MB under UV and visible light. GO was dispersed in aqueous solution with ultrasonication and then Zn powders were mixed with GO suspension. Zn powder was used as a reducing agent and a precursor of ZnO. Ye et al. [69] prepared CdS-graphene and CdS-carbon nanotube nanocomposites prepared with hydrothermal method and used these nanocomposites as photocatalysts for the hydrogen evolution and the degradation of methyl orange under visible-light irradiation.

4. Visible light photocatalytic activity of the graphene-semiconductor composites

In recent decades, graphene-semiconductor-assisted photocatalysts have attracted considerable attention in not only wastewater treatment but also antibacterial applications due to their high photosensitivity, low cost, and environment-friendly non-toxic nature [70]. Widely used photocatalyst semiconductors such as TiO_2, ZnO, CdS, CdSe, etc., do not fulfill all the practical needs as a photocatalytic material, since activation of these catalysts require high-energy UV light, resulting in low efficiency in the visible and near infrared regions. The low separation efficiency of electron-hole pairs and photo-corrosion are other principle problems that inhibit the photocatalytic activity of these semiconductors. In order to solve these limiting factors and enhance the visible light photocatalytic activity of the semiconductors, considerable efforts have been spent on the improvement of the photocatalytic activity by means of noble metal deposition [71], transition-metal doping [2], and semiconductor combination [72]. According to purely used semiconductor, better photocatalytic activities are obtained when the above listed methods are used. However, the recombination of electron-hole species in the semiconductors is still a problem for the photocatalytic applications.

Semiconductor-matrix system is another method to improve photocatalytic property due to impede the recombination of electron-hole species in the semiconductors by efficient electron transport matrices, such as carbon nanostructures or conductive polymer films, or by molecular electron delay semiconductor structures. Among these techniques, carbonaceous nanomaterials seems to be the most promising matrix system due to enhancement of the photocatalytic properties of semiconductors due to their unique structures that can bring attractive characteristic to photocatalyst materials. Photocatalyst and the delocalized conjugated carbonaceous materials have similar energy level providing an interface hybrid effect between these materials. This effect results in rapid charge separation and slow charge recombination in the electron-transfer process.

Among the various types of semiconductor materials, TiO_2 has provoked great interest as photocatalyst due to its excellent photocatalytic performance, high stability, low cost, and relatively low toxicity, in comparison to other semiconductor metal oxides [73]. Recently, TiO_2-carbonaceus-based composites attract great attention for their potential applications in air and water purification due to their outstanding photocatalytic activity. These types of composites can be divided into three main groups; carbon-doped TiO_2, TiO_2 fixed activated carbon, and carbon-coated TiO_2. Although unique advantages of carbon-based materials as composite material such as chemical stability and inertness in acid and basic media, unusual tunable structural, electronic and chemical properties [74, 75], the lack of reproducibility, and the weakening of the light intensity at catalyst surface have still remained as challenges

[76, 77]. Great photocatalytic activity under the visible light is another limiting factor that impedes the photocatalytic marketing.

Zhang et al. synthesized TiO_2 (P25)-graphene composite via a one-step hydrothermal method and investigated photocatalytic activity of the composites under the both UV and visible light [44]. They found that under both the light sources, the P25-graphene composite structure exhibited better photocatalytic degradation of the dye and enhanced charge separation and transportation properties than that of the pure P25 and P25-carbon nanoparticle structure. They also found that absorption spectra of pure P25 and P25-graphene composite structure shows a red shift to higher wavelength for composite structure exhibiting a photo-responding range at approximately 430–440 nm which corresponds to the violet-blue region in electromagnetic spectrum. Therefore, semiconductor-graphene composite structure offers a more efficient utilization of the solar spectrum under visible light irradiation and expected to assist its use in practical environmental protection issues. The improvement in the photocatalytic property of the P25-graphene composite structure can be attributed to its extended photo responding range by means of chemical bonds and selective adsorption of the aromatic dye on the catalyst. This noncovalent adsorption is driven by the π-π stacking between aromatic regions of the graphene and dye molecule. Similar adsorption has also been reported for the conjugation between aromatic molecules and carbon nanotubes [78].

Li et al. reported that graphene-CdS visible light driven photocatalysts composites act as the clean and renewable hydrogen production path through water splitting [58]. They found that under the visible light, pure CdS exhibited little photocatalytic activity due to the rapid recombination of CB-electrons and VB-holes. The photocatalytic activity of CdS semiconductor was improved by the graphene addition since larger surface area of graphene provides more active adsorption sites acting as photocatalytic reaction centers. Graphene also provides a reduction in the recombination probability of the photo-excited electron-hole pairs and an increasing the number of the charge carriers to form reactive species. However, there is saturation graphene amount in the visible light-induced photocatalytic activity due to "shielding effect." The high weight addition of graphene into composite structure reduces the light irradiation depth due to its opacity, thus and reaction medium restricts the efficiency of graphene in promoting the photocatalytic activity of graphene-semiconductor composites [44, 79, 80]. Due to intrinsic "shielding effect" leading to achieve an optimal photocatalytic activity enhancement with graphene addition into semiconductor material, generally 5 wt.% graphene is incorporated into composite structure.

For the investigation of visible light photocatalytic mechanism, Wang et al. reported synthesis of high-quality nanosized anatase ultra-thin TiO_2 nanosheets grown on graphene nanosheets with {001} facets via a simple one-pot solvothermal synthetic route [81]. They proposed that TiO_2/graphene nanocomposites show better visible light photocatalytic activity than pure TiO_2 and P25 due to prefer transferring of the photo-induced electrons via Ti—O—C between TiO_2 and C interaction. This electron transection delays the recombination of photo-induced charge carriers and extends the carrier lifetime, therefore provides to the enhancement of photocatalytic performance of the composite structure under the visible light. For this enhancement of the visible light photocatalytic activity of the composite structure, Wang et al.

proposed that the effective separation of photo-induced charges induced by TiO_2 nanosheets with exposed {001} facets and graphene plays an effective role.

5. Effects of doping on the visible light photocatalytic property of the graphene-semiconductor composites

Generally, in order to improve visible light-induced photocatalytic activity of the semiconductor-based, carbonaceous-based, and semiconductor-carbonaceous composite photocatalysts, the band gap energy of photocatalyst material should be reduced or split into several sub-gaps. Doping semiconductor-based photocatalyst material with various noble metals such as gold, palladium, and silver and transition metals such as cobalt, chrome, copper, iron, molybdenum, etc. become an exciting area in research [5, 82, 83]. Yildirim and Durucan reported that electronic band structure of the semiconductor ZnO nanoparticles can be tunable with substitutional incorporation of Cu ions into ZnO crystal lattice providing to the absorbing more photons [84]. Therefore, transition metal doping enables semiconductor photocatalyst to expand its photo response from the UV light region into the visible range and to decompose organic pollutants more effectively.

For the graphene-semiconductor-based composites, in order to improve the visible light-induced photocatalytic activity, the manipulation and optimization of charge carrier transfer across the interface between semiconductors and graphene is important [58]. Up to now, researchers have expressed an opinion that an intimate interfacial contact is a key aspect for the sufficient utilization of electron conductivity of graphene so enhancement of the photocatalytic activity of the graphene-semiconductor-based composites [85, 86]. However, tunable interfacial atomic charge carrier transfer pathways are also important for the photocatalytic activity of the composites. Atomic charge carrier transfer pathways can be controlled by means of the rational synergy interaction between incorporated individual components in graphene-semiconductor composites [3].

Zhang et al. reported a detailed study to investigate the effect of the various types of metal ion (Ca^{2+}, Cr^{3+}, Mn^{2+}, Fe^{2+}, Co^{2+}, Ni^{2+}, Cu^{2+}, and Zn^{2+}) addition into CdS which is a well-known II–VI semiconductor with suitable band gap (2.4 eV) matching well with the spectrum of sunlight [87]. Incorporation of the metal ions was performed by using the simple solvothermal method. They proposed a mechanism for the formation of the metal interfacial layer matrix between semiconductor CdS and graphene and found that incorporation of the small amount of metal ions into the interfacial layer matrix between the semiconductor and graphene considerably enhance the visible light-induced photocatalytic activity of graphene-semiconductor composites. Interfacial layer is maintained by the electrostatic attractive interaction between negatively charged oxygen atoms of the functional groups on the GO and positively charged metal ions. Furthermore, binding interaction between defect sites on the GO sheet and metal ions could also promote to more effectively attaching of metal ions onto the GO sheet. Zhang et al. also found that for the highest photocatalytic activity under the visible light irradiation, the optimal weight addition ratio of graphene is 5 wt.%. It is quite interesting to observe the best catalytic activity from 5 wt.% graphene addition since generally less than 5 wt.% graphene containing graphene-semiconductor composites exhibit a proper synergy interaction between graphene and the semiconductor due to the shielding effect of graphene. Graphene addition maintains the intimate interfacial contact between the semiconductor and graphene transfer charge carriers and expands the lifetime more effectively under visible light irradiation. Therefore, with the introducing small amount of metal ions into composite structure, the photocatalytic activity of the composite is considerably improved. Under visible light irradiation, this composite structure can be used anaerobic reduction of nitro compound and the aerobic selective oxidation of alcohol.

In order to improve the visible light-induced photocatalytic activity, doping was found to be helpful due to tailoring the electronic properties of the composites. Mechanism of the organic dye degradation by the semiconductor catalyst is based on two photocatalytic routes for the excitation under visible-light irradiation. A schematic representation of the photocatalytic reaction mechanism for the Ag-doped TiO_2-rGO composite structure is shown in **Figure 2**. The first photocatalytic route is based on the excitation of the semiconductor TiO_2. Light irradiation causes to electron excitation forming photogenerated electron in the conduction band (CB)/hole in the valence band (VB). Photoinduced electron transfer could take place from CB of TiO_2 (−4.2 eV) to work function of the graphene oxide (−4.42 eV) [88]. Furthermore, the work function of the Ag nanoparticles (−4.74 eV) is located at a lower energy level than that of GO [89]. Therefore, photoinduced electrons could easily move from GO to Ag nanoparticle surface and chemical reactions with the dye molecules occur as photogenerated charges move to the particle surface. The recombination of photoinduced electrons and holes is minimized, thus reactive oxygen radicals ($^{\bullet}O_2^-$) is formed as a reaction of these electrons with adsorbed oxidants (usually O_2). At the same time, the holes in the VB could react with H_2O/OH^- to generate hydroxyl radicals ($^{\bullet}OH$) which contribute to dye degradation.

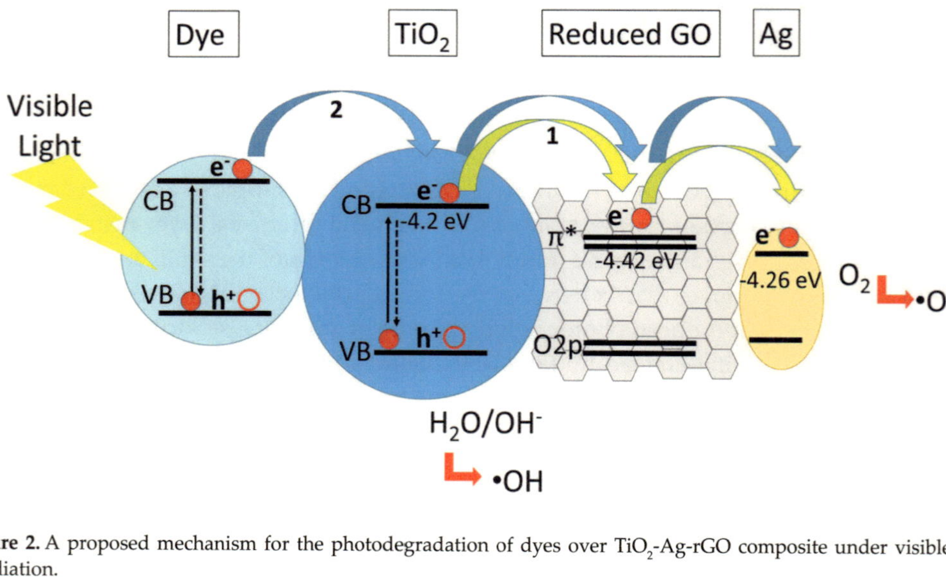

Figure 2. A proposed mechanism for the photodegradation of dyes over TiO_2-Ag-rGO composite under visible light irradiation.

The second photocatalytic route includes the excitation of dye under visible-light irradiation. In this way, photogenerated electrons excited from dye could transfer to first TiO_2, then to graphene oxide, and to lastly Ag nanoparticle surface as shown in **Figure 2**. Therefore, dye molecules provide also to an improvement of the photocatalytic ability by means of acting as a visible light sensitizer and offering a self-degradation. For the both routes, while reaction of the excited electrons and dissolved oxygen molecules produces oxygen peroxide radicals, reaction between the positive charged hole and the hydroxide ion derived from water forms hydroxyl radicals. As a result of these reactions, an enhancement of the photocatalytic activity of GO-TiO_2-Ag nanocomposites is expected.

Lu et al. reported that manganese oxides supported on TiO_2-graphene nanocomposite catalysts [90]. MnO_X active component was incorporated into TiO_2-graphene nanocomposites prepared with the sol-gel method and contained different amount of graphene (0–2 wt.%). These MnOx incorporated TiO_2-graphene catalysts exhibit excellent structure and electrical properties, which favored the catalytic reaction. Due to different oxide states of Mn (MnO, MnO_2, Mn_2O_3, and nonstoichiometric MnO_X/Mn) Mn incorporation provides multiple valence states which supports electron transfer and so redox reaction on the surface of the composite catalyst.

6. Conclusions

In recent decades, photocatalysis is attracting more and more attention due to their potential for solving environmental problems originated from organic dye compounds which cause to serious detrimental effects for living ecosystems as a result of water pollution. Up to now, many research papers and review articles are dedicated to development of the photocatalytic materials. Semiconductor metal oxide especially TiO_2 and ZnO have been used as two most preferred materials widely used in the removal of toxic or hazardous organic pollutants. However, due to their wide band gap, semiconductor-based photocatalysts can only absorb UV light comprising for less than 10% of solar irradiation. Therefore, expanding the absorption spectral range to exploit the more abundant source of light energy and even poor illumination of interior lighting coming from the sun is critical in development of the photocatalyst. According to literature, the photocatalytic activity of semiconductor materials can be improved by the graphene addition since graphene with larger surface area provides more active adsorption sites acting as photocatalytic reaction centers. Graphene also provides a reduction in the recombination probability of the photo-excited electron-hole pairs and an increasing the number of the charge carriers to form reactive species. For this purpose, graphene-semiconductor-based composite structure is one of the most popular approach due to their outstanding/combining advantages such as theoretically large specific surface area and high intrinsic electron mobility of graphene at room temperature, low cost, innocuous nature, ease of availability/processing, and suitability of their structure to create composites with each other.

In literature, a wide variety of the growth methods have been reported such as sol-gel, solution mixing, *in situ* growth, hydrothermal growth, and solvothermal method. Among these methods, solution mixing technique is one of the most common and easiest the method for the fabrication of graphene-semiconductor composite structures which are synthesized by using the suspensions of semiconductor nanoparticles and graphene. For the graphene-semiconductor-based composites, to enhance the visible light-induced photocatalytic activity, doping was found to be helpful due to tailoring the electronic properties of the composites providing the manipulation and optimization of charge carrier transfer across the interface between semiconductors and graphene and a reduction in the recombination probability of the photo-excited electron-hole pairs. This chapter of the book is a review of the studies for the development of highly photocatalytically active graphene-semiconductor composite under visible light irradiation.

Author details

Ozlem Altintas Yildirim[1]* and Teoman Ozturk[2]

*Address all correspondence to: ozlemaltintas@gmail.com

1 Department of Metallurgical and Materials Engineering, Selcuk University, Konya, Turkey

2 Department of Physics, Selcuk University, Konya, Turkey

References

[1] Sonune A, Ghate R. Developments in wastewater treatment methods. Desalination. 2004;**167**:55-63

[2] Yildirim OA, Arslan H, Sönmezoğlu S. Facile synthesis of cobalt-doped zinc oxide thin films for highly efficient visible light photocatalysts. Applied Surface Science. 2016;**390**: 111-121

[3] Zhang N, Zhang Y, Xu Y-J. Recent progress on graphene-based photocatalysts: Current status and future perspectives. Nanoscale. 2012;**4**:5792-5813

[4] Upadhyay RK, Soin N, Roy SS. Role of graphene/metal oxide composites as photocatalysts, adsorbents and disinfectants in water treatment: A review. RSC Advances. 2014;**4**:3823-3851

[5] Yıldırım ÖA, Unalan HE, Durucan C. Highly efficient room temperature synthesis of silver-doped zinc oxide (ZnO:Ag) nanoparticles: Structural, optical, and photocatalytic properties. Journal of the American Ceramic Society. 2013;**96**:766-773

[6] An X, Jimmy CY. Graphene-based photocatalytic composites. RSC Advances. 2011;**1**: 1426-1434

[7] Novoselov KS, Geim AK, Morozov SV, Jiang D, Zhang Y, Dubonos SV, Grigorieva IV, Firsov AA. Electric field effect in atomically thin carbon films. Science. 2004;**306**:666-669

[8] Mayorov AS, Gorbachev RV, Morozov SV, Britnell L, Jalil R, Ponomarenko LA, Blake P, Novoselov KS, Watanabe K, Taniguchi T. Micrometer-scale ballistic transport in encapsulated graphene at room temperature. Nano Letters. 2011;**11**:2396-2399

[9] Stoller MD, Park S, Zhu Y, An J, Ruoff RS. Graphene-based ultracapacitors. Nano Letters. 2008;**8**:3498-3502

[10] Nair RR, Blake P, Grigorenko AN, Novoselov KS, Booth TJ, Stauber T, Peres NM, Geim AK. Fine structure constant defines visual transparency of graphene. Science. 2008;**320**:1308-1308

[11] Lee C, Wei X, Kysar JW, Hone J. Measurement of the elastic properties and intrinsic strength of monolayer graphene. Science. 2008;**321**:385-388

[12] Ghosh S, Calizo I, Teweldebrhan D, Pokatilov EP, Nika DL, Balandin AA, Bao W, Miao F, Lau CN. Extremely high thermal conductivity of graphene: Prospects for thermal management applications in nanoelectronic circuits. Applied Physics Letters. 2008;**92**:151911

[13] Novoselov KS, Jiang Z, Zhang Y, Morozov SV, Stormer HL, Zeitler U, Maan JC, Boebinger GS, Kim P, Geim AK. Room-temperature quantum hall effect in graphene. Science. 2007;**315**:1379-1379

[14] El-Kady MF, Strong V, Dubin S, Kaner RB. Laser scribing of high-performance and flexible graphene-based electrochemical capacitors. Science. 2012;**335**:1326-1330

[15] Fu Y, Wang X. Magnetically separable $ZnFe_2O_4$–graphene catalyst and its high photocatalytic performance under visible light irradiation. Industrial and Engineering Chemistry Research. 2011;**50**:7210-7218

[16] Li W, Geng X, Guo Y, Rong J, Gong Y, Wu L, Zhang X, Li P, Xu J, Cheng G. Reduced graphene oxide electrically contacted graphene sensor for highly sensitive nitric oxide detection. ACS Nano. 2011;**5**:6955-6961

[17] Wakabayashi, Katsunori. Electronic Properties of Graphene Nanoribbons. Graphene Nanoelectronics. Springer Berlin Heidelberg, 2011. 277-299

[18] Bonaccorso F, Colombo L, Yu G, Stoller M, Tozzini V, Ferrari AC, Ruoff RS, Pellegrini V. Graphene, related two-dimensional crystals, and hybrid systems for energy conversion and storage. Science. 2015;**347**:1246501

[19] Niu P, Zhang L, Liu G, Cheng HM. Graphene-like carbon nitride nanosheets for improved photocatalytic activities. Advanced Functional Materials. 2012;**22**:4763-4770

[20] Kroto HW, Heath JR, O'Brien SC, Curl RF, Smalley RE. C 60: Buckminsterfullerene. Nature. 1985;**318**:162-163

[21] Berger C, Song Z, Li T, Li X, Ogbazghi AY, Feng R, Dai Z, Marchenkov AN, Conrad EH, First PN. Ultrathin epitaxial graphite: 2D electron gas properties and a route toward graphene-based nanoelectronics. The Journal of Physical Chemistry B. 2004;**108**:19912-19916

[22] Reina A, Jia X, Ho J, Nezich D, Son H, Bulovic V, Dresselhaus MS, Kong J. Large area, few-layer graphene films on arbitrary substrates by chemical vapor deposition. Nano Letters. 2008;**9**:30-35

[23] Kim KS, Zhao Y, Jang H, Lee SY, Kim JM, Kim KS, Ahn J-H, Kim P, Choi J-Y, Hong BH. Large-scale pattern growth of graphene films for stretchable transparent electrodes. Nature. 2009;**457**:706-710

[24] Stankovich S, Piner RD, Chen X, Wu N, Nguyen ST, Ruoff RS. Stable aqueous dispersions of graphitic nanoplatelets via the reduction of exfoliated graphite oxide in the presence of poly (sodium 4-styrenesulfonate). Journal of Materials Chemistry. 2006;**16**:155-158

[25] Park S, Ruoff RS. Chemical methods for the production of graphenes. Nature Nanotechnology. 2009;**4**:217-224

[26] Stankovich S, Dikin DA, Dommett GH, Kohlhaas KM, Zimney EJ, Stach EA, Piner RD, Nguyen ST, Ruoff RS. Graphene-based composite materials. Nature. 2006;**442**:282-286

[27] Stankovich S, Dikin DA, Piner RD, Kohlhaas KA, Kleinhammes A, Jia Y, Wu Y, Nguyen ST, Ruoff RS. Synthesis of graphene-based nanosheets via chemical reduction of exfoliated graphite oxide. Carbon. 2007;**45**:1558-1565

[28] Hummers WS, Offeman RE. Preparation of graphitic oxide. Journal of the American Chemical Society. 1958;**80**:1339-1339

[29] Eda G, Chhowalla M. Chemically derived graphene oxide: Towards large-area thin-film electronics and optoelectronics. Advanced Materials. 2010;**22**:2392-2415

[30] Dreyer DR, Park S, Bielawski CW, Ruoff RS. The chemistry of graphene oxide. Chemical Society Reviews. 2010;**39**:228-240

[31] Pei S, Cheng H-M. The reduction of graphene oxide. Carbon. 2012;**50**:3210-3228

[32] Wang G, Yang J, Park J, Gou X, Wang B, Liu H, Yao J. Facile synthesis and characterization of graphene nanosheets. The Journal of Physical Chemistry C. 2008;**112**:8192-8195

[33] Shin HJ, Kim KK, Benayad A, Yoon SM, Park HK, Jung IS, Jin MH, Jeong HK, Kim JM, Choi JY. Efficient reduction of graphite oxide by sodium borohydride and its effect on electrical conductance. Advanced Functional Materials. 2009;**19**:1987-1992

[34] Moon IK, Lee J, Ruoff RS, Lee H. Reduced graphene oxide by chemical graphitization. Nature Communications. 2010;**1**:73

[35] Schniepp HC, Li J-L, McAllister MJ, Sai H, Herrera-Alonso M, Adamson DH, Prud'homme RK, Car R, Saville DA, Aksay IA. Functionalized single graphene sheets derived from splitting graphite oxide. The Journal of Physical Chemistry B. 2006;**110**:8535-8539

[36] McAllister MJ, Li J-L, Adamson DH, Schniepp HC, Abdala AA, Liu J, Herrera-Alonso M, Milius DL, Car R, Prud'homme RK. Single sheet functionalized graphene by oxidation and thermal expansion of graphite. Chemistry of Materials. 2007;**19**:4396-4404

[37] Yang D, Velamakanni A, Bozoklu G, Park S, Stoller M, Piner RD, Stankovich S, Jung I, Field DA, Ventrice CA. Chemical analysis of graphene oxide films after heat and chemical treatments by X-ray photoelectron and micro-Raman spectroscopy. Carbon. 2009;**47**:145-152

[38] Toh SY, Loh KS, Kamarudin SK, Daud WRW. Graphene production via electrochemical reduction of graphene oxide: Synthesis and characterisation. Chemical Engineering Journal. 2014;**251**:422-434

[39] Zhou M, Wang Y, Zhai Y, Zhai J, Ren W, Wang F, Dong S. Controlled synthesis of large-area and patterned electrochemically reduced graphene oxide films. Chemistry—A European Journal. 2009;**15**:6116-6120

[40] Zedan AF, Moussa S, Terner J, Atkinson G, El-Shall MS. Ultrasmall gold nanoparticles anchored to graphene and enhanced photothermal effects by laser irradiation of gold nanostructures in graphene oxide solutions. ACS Nano. 2012;**7**:627-636

[41] Low J, Yu J, Li Q, Cheng B. Enhanced visible-light photocatalytic activity of plasmonic Ag and graphene co-modified Bi_2WO_6 nanosheets. Physical Chemistry Chemical Physics. 2014;**16**:1111-1120

[42] Fujishima A, Honda K. Electrochemical photolysis of water at a semiconductor electrode. Nature. 1972;**238**:37-38

[43] Williams G, Seger B, Kamat PV. TiO_2-graphene nanocomposites. UV-assisted photocatalytic reduction of graphene oxide. ACS Nano. 2008;**2**:1487-1491

[44] Zhang H, Lv X, Li Y, Wang Y, Li J. P25-graphene composite as a high performance photocatalyst. ACS Nano. 2010;**4**:380-386

[45] Liu Q, Liu Z, Zhang X, Yang L, Zhang N, Pan G, Yin S, Chen Y, Wei J. Polymer photovoltaic cells based on solution-processable graphene and P3HT. Advanced Functional Materials. 2009;**19**:894-904

[46] Tan LL, Chai SP, Mohamed AR. Synthesis and applications of graphene-based TiO_2 photocatalysts. ChemSusChem. 2012;**5**:1868-1882

[47] Liu J, Wang Z, Liu L, Chen W. Reduced graphene oxide as capturer of dyes and electrons during photocatalysis: Surface wrapping and capture promoted efficiency. Physical Chemistry Chemical Physics. 2011;**13**:13216-13221

[48] Yu J-G, Yu H-G, Cheng B, Zhao X-J, Yu JC, Ho W-K. The effect of calcination temperature on the surface microstructure and photocatalytic activity of TiO_2 thin films prepared by liquid phase deposition. The Journal of Physical Chemistry B. 2003;**107**:13871-13879

[49] Sakthivel S, Shankar M, Palanichamy M, Arabindoo B, Bahnemann D, Murugesan V. Enhancement of photocatalytic activity by metal deposition: Characterisation and photonic efficiency of Pt, Au and Pd deposited on TiO_2 catalyst. Water Research. 2004;**38**:3001-3008

[50] In S, Orlov A, Berg R, García F, Pedrosa-Jimenez S, Tikhov MS, Wright DS, Lambert RM. Effective visible light-activated B-doped and B, N-codoped TiO_2 photocatalysts. Journal of the American Chemical Society. 2007;**129**:13790-13791

[51] Subramanian V, Wolf E, Kamat PV. Semiconductor–metal composite nanostructures. To what extent do metal nanoparticles improve the photocatalytic activity of TiO_2 films? The Journal of Physical Chemistry B. 2001;**105**:11439-11446

[52] Marcì G, Augugliaro V, López-Muñoz MJ, Martín C, Palmisano L, Rives V, Schiavello M, Tilley RJ, Venezia AM. Preparation characterization and photocatalytic activity of polycrystalline ZnO/TiO_2 systems. 1. Surface and bulk characterization. The Journal of Physical Chemistry B. 2001;**105**:1026-1032

[53] Ozgur U, Alivov YI, Liu C, Teke A, Reshchikov MA, Dogan S, Avrutin V, Cho SJ, Morkoc H. A comprehensive review of ZnO materials and devices. Journal of Applied Physics. 2005;**98**:041301

[54] Williams G, Kamat PV. Graphene–semiconductor nanocomposites: Excited-state interactions between ZnO nanoparticles and graphene oxide. Langmuir. 2009;**25**:13869-13873

[55] Xu T, Zhang L, Cheng H, Zhu Y. Significantly enhanced photocatalytic performance of ZnO via graphene hybridization and the mechanism study. Applied Catalysis B: Environmental. 2011;**101**:382-387

[56] Seema H, Kemp KC, Chandra V, Kim KS. Graphene–SnO_2 composites for highly efficient photocatalytic degradation of methylene blue under sunlight. Nanotechnology. 2012;**23**:355705

[57] Ng YH, Iwase A, Kudo A, Amal R. Reducing graphene oxide on a visible-light $BiVO_4$ photocatalyst for an enhanced photoelectrochemical water splitting. The Journal of Physical Chemistry Letters. 2010;**1**:2607-2612

[58] Li Q, Guo B, Yu J, Ran J, Zhang B, Yan H, Gong JR. Highly efficient visible-light-driven photocatalytic hydrogen production of CdS-cluster-decorated graphene nanosheets. Journal of the American Chemical Society. 2011;**133**:10878-10884

[59] Jia L, Wang D-H, Huang Y-X, Xu A-W, Yu H-Q. Highly durable N-doped graphene/CdS nanocomposites with enhanced photocatalytic hydrogen evolution from water under visible light irradiation. The Journal of Physical Chemistry C. 2011;**115**:11466-11473

[60] Zhang X-Y, Li H-P, Cui X-L, Lin Y. Graphene/TiO_2 nanocomposites: Synthesis, characterization and application in hydrogen evolution from water photocatalytic splitting. Journal of Materials Chemistry. 2010;**20**:2801-2806

[61] Liu S, Sun H, Liu S, Wang S. Graphene facilitated visible light photodegradation of methylene blue over titanium dioxide photocatalysts. Chemical Engineering Journal. 2013;**214**:298-303

[62] Paek S-M, Yoo E, Honma I. Enhanced cyclic performance and lithium storage capacity of SnO_2/graphene nanoporous electrodes with three-dimensionally delaminated flexible structure. Nano Letters. 2009;**9**:72-75

[63] Bell NJ, Ng YH, Du A, Coster H, Smith SC, Amal R. Understanding the enhancement in photoelectrochemical properties of photocatalytically prepared TiO_2-reduced graphene oxide composite. The Journal of Physical Chemistry C. 2011;**115**:6004-6009

[64] Akhavan O, Ghaderi E. Photocatalytic reduction of graphene oxide nanosheets on TiO_2 thin film for photoinactivation of bacteria in solar light irradiation. The Journal of Physical Chemistry C. 2009;**113**:20214-20220

[65] Lambert TN, Chavez CA, Hernandez-Sanchez B, Lu P, Bell NS, Ambrosini A, Friedman T, Boyle TJ, Wheeler DR, Huber DL. Synthesis and characterization of titania–graphene nanocomposites. The Journal of Physical Chemistry C. 2009;**113**:19812-19823

[66] Zhang J, Xiong Z, Zhao XS. Graphene-metal-oxide composites for the degradation of dyes under visible light irradiation. Journal of Materials Chemistry. 2011;**21**:3634-3640

[67] Li B, Cao H. ZnO–graphene composite with enhanced performance for the removal of dye from water. Journal of Materials Chemistry. 2011;**21**:3346-3349

[68] Liu S, Sun H, Suvorova A, Wang S. One-pot hydrothermal synthesis of ZnO-reduced graphene oxide composites using Zn powders for enhanced photocatalysis. Chemical Engineering Journal. 2013;**229**:533-539

[69] Ye A, Fan W, Zhang Q, Deng W, Wang Y. CdS-graphene and CdS-CNT nanocomposites as visible-light photocatalysts for hydrogen evolution and organic dye degradation. Catalysis Science & Technology. 2012;**2**:969-978

[70] Li X, Yu J, Jaroniec M. Hierarchical photocatalysts. Chemical Society Reviews. 2016;**45**:2603-2636

[71] Seery MK, George R, Floris P, Pillai SC. Silver doped titanium dioxide nanomaterials for enhanced visible light photocatalysis. Journal of Photochemistry and Photobiology A: Chemistry. 2007;**189**:258-263

[72] Cho S, Jang J-W, Kim J, Lee JS, Choi W, Lee K-H. Three-dimensional type II ZnO/ZnSe heterostructures and their visible light photocatalytic activities. Langmuir. 2011; **27**:10243-10250

[73] Morales-Torres S, Pastrana-Martínez LM, Figueiredo JL, Faria JL, Silva AMT. Design of graphene-based TiO_2 photocatalysts—A review. Environmental Science and Pollution Research. 2012;**19**:3676-3687

[74] Sampaio MJ, Silva CG, Marques RR, Silva AM, Faria JL. Carbon nanotube–TiO_2 thin films for photocatalytic applications. Catalysis Today. 2011;**161**:91-96

[75] Wang F, Zhang K. Physicochemical and photocatalytic activities of self-assembling TiO_2 nanoparticles on nanocarbons surface. Current Applied Physics. 2012;**12**:346-352

[76] Woan K, Pyrgiotakis G, Sigmund W. Photocatalytic carbon-nanotube–TiO_2 composites. Advanced Materials. 2009;**21**:2233-2239

[77] Inagaki M, Kojin F, Tryba B, Toyoda M. Carbon-coated anatase: The role of the carbon layer for photocatalytic performance. Carbon. 2005;**43**:1652-1659

[78] Liu Z, Robinson JT, Sun X, Dai H. PEGylated nanographene oxide for delivery of water-insoluble cancer drugs. Journal of the American Chemical Society. 2008;**130**:10876-10877

[79] Huang X, Qi X, Boey F, Zhang H. Graphene-based composites. Chemical Society Reviews. 2012;**41**:666-686

[80] Zhang N, Zhang Y, Yang M-Q, Tang Z-R, Xu Y-J. A critical and benchmark comparison on graphene-, carbon nanotube-, and fullerene-semiconductor nanocomposites as visible light photocatalysts for selective oxidation. Journal of Catalysis. 2013;**299**:210-221

[81] Wang W-S, Wang D-H, Qu W-G, Lu L-Q, Xu A-W. Large ultrathin anatase TiO_2 nanosheets with exposed {001} facets on graphene for enhanced visible light photocatalytic activity. The Journal of Physical Chemistry C. 2012;**116**:19893-19901

[82] Ahmad M, Yingying S, Nisar A, Sun H, Shen W, Wei M, Zhu J. Synthesis of hierarchical flower-like ZnO nanostructures and their functionalization by Au nanoparticles for improved photocatalytic and high performance Li-ion battery anodes. Journal of Materials Chemistry. 2011;**21**:7723-7729

[83] Georgekutty R, Seery MK, Pillai SC. A highly efficient Ag-ZnO photocatalyst: Synthesis, properties, and mechanism. The Journal of Physical Chemistry C. 2008;**112**:13563-13570

[84] Altıntaş Yıldırım Ö, Durucan C. Room temperature synthesis of Cu incorporated ZnO nanoparticles with room temperature ferromagnetic activity: Structural, optical and magnetic characterization. Ceramics International. 2016;**42**:3229-3238

[85] Fan W, Lai Q, Zhang Q, Wang Y. Nanocomposites of TiO_2 and reduced graphene oxide as efficient photocatalysts for hydrogen evolution. The Journal of Physical Chemistry C. 2011;**115**:10694-10701

[86] Xiang Q, Yu J, Jaroniec M. Preparation and enhanced visible-light photocatalytic H_2-production activity of graphene/C_3N_4 composites. The Journal of Physical Chemistry C. 2011;**115**:7355-7363

[87] Zhang N, Yang M-Q, Tang Z-R, Xu Y-J. Toward improving the graphene–semiconductor composite photoactivity via the addition of metal ions as generic interfacial mediator. ACS Nano. 2014;**8**:623-633

[88] Liu L, Bai H, Liu J, Sun DD. Multifunctional graphene oxide-TiO_2-Ag nanocomposites for high performance water disinfection and decontamination under solar irradiation. Journal of Hazardous Materials. 2013;**261**:214-223

[89] Yoo D-H, Cuong TV, Luan VH, Khoa NT, Kim EJ, Hur SH, Hahn SH. Photocatalytic performance of a Ag/ZnO/CCG multidimensional heterostructure prepared by a solution-based method. The Journal of Physical Chemistry C. 2012;**116**:7180-7184

[90] Lu X, Song C, Chang C-C, Teng Y, Tong Z, Tang X. Manganese oxides supported on TiO_2–graphene nanocomposite catalysts for selective catalytic reduction of NO_X with NH_3 at low temperature. Industrial and Engineering Chemistry Research. 2014;**53**:11601-11610

Chapter 6

Role of Graphene in Photocatalytic Solar Fuel Generation

Babak Adeli and Fariborz Taghipour

Additional information is available at the end of the chapter

http://dx.doi.org/10.5772/intechopen.72623

Abstract

One of the most promising methods for conversion and storage of solar energy is in the form of the chemical bonds of an energy carrier, such as hydrogen or light hydrocarbons. However, the traditional methods to harness and store solar energy are simply too expensive to be implemented on a large scale. It has been documented that the recombination of photo-induced charge carriers is the greatest source of inefficiency in photocatalytic systems. In the last decade, graphene derivatives and their functionalized nanostructures were extensively utilized for various roles to improve the efficiency of photocatalytic solar fuel generation. These include photocatalyst/redox active sites via band gap and defect density engineering, charge acceptor due to their excellent carrier mobility, a solid-state charge mediator by electronic band alignment, and light absorber by taking advantage of their photoluminescence characteristics at the nanoscale. This chapter aims to provide an authoritative and in-depth review on the properties and application of graphene derivatives, as well as the recent advances in the design of graphene-based photocatalytic systems. The knowledge extracted from the presented materials can be applied to other applications dealing with surface chemistry, interfacial science, and optoelectronic device fabrication.

Keywords: solar fuel generation, graphene, photocatalyst, water splitting, CO_2 reduction

1. Introduction

Direct production of fuels from sunlight is an attractive route to address the energy crisis facing humanity in the twenty-first century, because it inherently provides a method for extracting energy during the night and for cost-effectively dispatching and distributing energy in the existing infrastructure for use in the residential, industrial, and transportation sectors [1]. That places the photocatalytic splitting of water and the conversion of carbon dioxide (CO_2) to light hydrocarbons, driven solely by sunlight, among the most promising approaches. These were studied extensively in the last decade [2]. However, visible-light water splitting and CO_2 reduction are inherently associated with inefficiencies and complicated processes. For instance, the possible products of these processes may include H_2, HCOOH, HCHO, CH_3OH, CO, and CH_4, which are selective and dependent on many competing factors (e.g., reaction kinetics, redox potentials of photo-induced charge carriers, morphology, crystalline structure, exposed crystalline facets and the surface properties of the utilized photocatalyst, and redox active sites) [3]. In addition, the energy levels of the photo-induced charges are relative to the electronic band structure of the employed semiconductor; thus, the desired photocatalyst must possess a matching molecular orbital structure corresponding to the redox potentials of the reaction. Electrons contain the energy of the lowest unoccupied molecular orbital (LUMO) and holes pose the potential energy of the highest occupied molecular orbital (HOMO). These energy levels are also known as the bottom of the semiconductor's conduction band (CB) and the top of the semiconductor's valence band (VB) and are shown in **Figure 1**. For a photocatalytic reaction to proceed, the photo-induced charges must pose a suitable energy level corresponding to redox potential of the reaction. Taking the solar water-splitting reaction as an example, the photo-excited electrons and the holes must contain more negative and more positive energy compared to the water reduction (0 eV vs. NHE) and the water oxidation potential (1.23 eV vs. NHE), respectively. Upon photo-excitation and the generation of photo-induced electrons and holes, the charge carriers can reach to the electrolyte and participate in redox reactions at CB and VB, as demonstrated in **Figure 1**.

Despite the efforts in utilizing metal oxide, sulfide, and nitride photocatalysts, as well as their binary and ternary solid solutions, the efficiencies of the solar fuel generation processes have

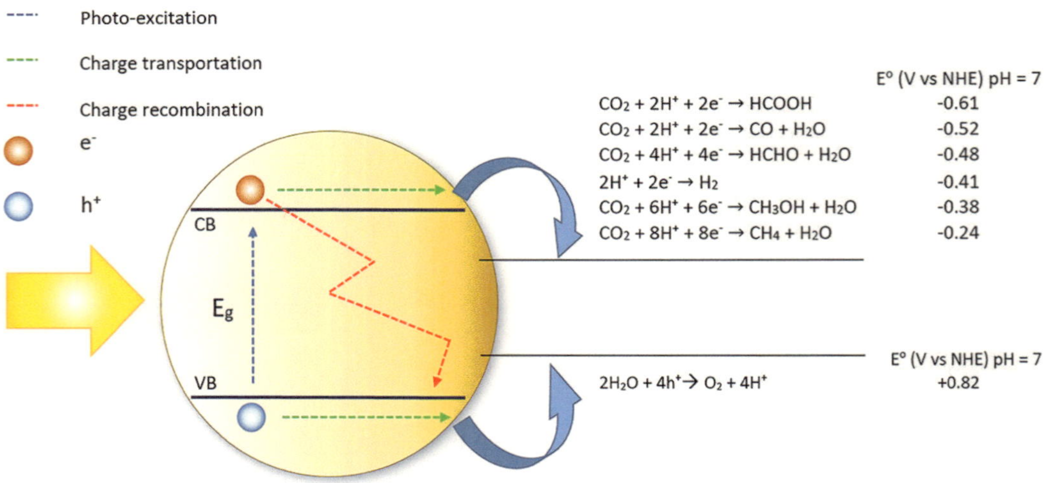

Figure 1. Band gap energy (E_g), valence band (VB), and conduction band (CB) potential of semiconductors. The figure also shows processes (photo-excitation, charge transportation, and charge recombination) that occur upon striking the surface of a semiconductor by a light photon with energy greater than the semiconductor's band gap. Once photo-induced charges reach the surface of the semiconductor, several possible products can proceed (i.e., water splitting and CO_2 conversion). The potential of the redox reactions is denoted by E°.

remained too low for them to be feasibly commercialized. Such inefficiencies are primarily attributed in the recombination of photo-induced charges, displayed by red dashed-line in **Figure 1**, which occurs mainly at the grain boundaries and the crystalline defects within the bulk of photocatalyst, where the diffusion path of charge carriers is considerable, and/or the density redox active sites are not sufficient [4]. In the last few years, various strategies—emphasizing the nanoscale morphology and exposed crystallographic facets, as well as increasing the specific surface area and the number of redox active sites—have led to outstanding improvements. However, these are not enough to put the commercialization of solar water splitting and CO_2 reduction in sight. Moreover, the abundance of photocatalyst materials and fabrication routes of advance structures has hindered further consideration of some candidate materials.

In the last decade, carbon-based materials, such as carbon nanotubes (CNTs), graphene, graphene oxide, carbon quantum dots, carbon fibers, activated carbon, and carbon black, have been the focus of intense research, owing to their peculiar characteristics, such as tunable electronic band structure, ultra-high specific surface area, tailored crystalline structure, and reactive crystallographic facets. Among them, graphene derivatives have grasped researchers' attention, due to their effectiveness as redox active sites, tunable defect-density active sites, short carrier's diffusion paths, and high electron mobility, as well as efficient light harvesting within their two-dimensional (2D) crystallography.

Over 18,000 articles related to graphene were published from 2004 to 2014. In the last decade, due to the advances in materials sciences and nanotechnology, tailoring the optical, structural, and electrochemical characteristics of graphene-based photocatalysts at the nanoscale toward quantum efficiency (QE) improvement has been extensively studied. This chapter aims to present the recent advances in the application of graphene-based materials in solar fuel generation via water splitting and CO_2 conversion. Readers are encouraged to reach out to comprehensive review articles previously published on topics related to the subject of this chapter [3, 5–9].

2. Graphene derivative materials

Dating back to October 2004, a revolution in science and technology was triggered when Novoselov et al. [10] had prepared stable 2D sheets of carbon atoms at ambient conditions, the so-called graphene nanosheets. Graphene, as an allotrope of carbon, is an isolated monolayer sheet containing atoms that are tightly packed into an sp^2 honeycomb lattice hybridized C–C bond with a π-electron cloud and is considered as one of the most important materials in the current century. Graphene soon became one of the attractive components in photonic device fabrication, fuel conversion, fuel storage, environment, sensing, and catalysis, owing to its outstanding mechanical, thermal, optical, and electrical properties. Photocatalytic applications, highly conductive graphene nanosheets (2D) and quantum dots (zero-dimensional) with a massive surface area and ultra-active catalytic facets, particularly on the dangling crystallography edges, are excellent materials for hybridization with prominent photocatalysts to enhance the separation of photo-excited charges and the active surface area for the redox reactions.

Early studies showed that monolayer graphene can be successfully isolated and studied; while a 2D crystalline graphene nanosheet is known to be thermodynamically unstable, its properties are not yet well known [11]. Graphene is a zero band gap semimetal with a small overlap between its HOMO and LUMO [12–14]. In fact, the electrical, mechanical, optical, and thermal properties of graphene are very similar to those of single-walled carbon nanotubes (SWCNTs), while it can be prepared at significantly lower cost [11]. In particular, graphene's massive theoretical specific surface area (~2600 $m^2\ g^{-1}$ [15]), high mobility of charge carriers (~10,000 $cm^2\ V^{-1}\ s^{-1}$ at room temperature, approaching to 200,000 $cm^2\ V^{-1}\ s^{-1}$ for lower carrier densities and temperatures [16]), plus its excellent thermal conductivity (3000–5000 $W\ m^{-1}\ K^{-1}$ [17]), 97.7% optical transmittance, and <0.1% reflectance [18] (monolayer graphene nanosheet) make it an excellent choice for light harvesting and the fabrication of energy conversion devices [15, 16, 19].

Within a few years, graphene-derivative materials, such as graphene oxide (GO), reduced graphene oxide (rGO), nanoribbons, quantum dots (QDs), and their functionalized nanostructures, demonstrated even more exciting characteristics. GO is a 2D carbon nanomaterial with many merits, such as low manufacturing cost, facile mass production, fascinating chemistry, and remarkable semiconducting behavior [5, 20–25]. The oxygen-containing functional groups on the surface of GO make it readily dispersed in aqueous solutions and effectively interact with other organic and inorganic compounds, as well as ionic species [26]. The surface of GO nanosheets is mainly decorated with epoxide (═O) and hydroxyl (—OH) groups, while small composition of carbonyl (—C═O) and carboxyl (—COOH) groups are linked to the nanosheet's edges [27]. These oxygen functionalities allow GO and rGO to interact with a wide range of precursors and structures through noncovalent, covalent, and/or ionic interactions [15, 18]. In addition, various densities of sp^3 hybridization in the GO structure create a wide range of interesting characteristics. Unlike sp^2 hybridization, which is attributed to the bonding of carbon atoms to the neighboring carbon atoms (which are not connected with hydroxy or epoxy groups) or oxygen in the form of carbonyl or carboxyl groups, sp^3 hybridization forms when carbon atoms are bonded to epoxy or hydroxyl groups [28–30]. Density functional theory (DFT) studies, in agreement with experimental observations, confirm the role of oxygen functional groups on the optical properties of graphene over a wide range [29–31]. An intense blue shift in the electron energy loss spectrum (EELS) of GO was observed when the concentration of epoxy and hydroxyl functional groups increased [29, 30]. Interestingly, an increase in the density of carbonyl groups affects the GO EELS; as such, at O/C ~37.5% a red shift about 1.0 eV compared to the pristine graphene is observed [29]. Therefore, the density of surface functionalities can be precisely tuned to control the optical and electrochemical properties of graphene derivatives over a wide range.

QDs, a multilayer zero-dimensional structure, contain sp^2 hybridized honeycomb carbon, have attracted enormous attention, and have been extensively studied for a wide range of applications—energy conversion in particular [32, 33]. Within their ultra-small sizes (typically less than 100 nm), their optical and electronic properties can be tailored, and a well-defined band gap can be formed [34]. Therefore, graphene QDs are like highly crystalline inorganic semiconductor QDs with superior physically and chemically reactive facets [32].

3. Role of graphene in solar fuel generation

The improvements in the efficiencies of graphene-based photocatalyst are likely the outcome of these various functionalities and cannot be attributed to one individual phenomenon. In the following sections, the role of graphene derivatives in visible light-driven fuel generation is discussed, and examples from literature are presented.

3.1. Graphene as photocatalyst

There are a handful of materials that can carry out visible light solar fuel generation reaction, exhibiting properties such as short band gap to harvest visible light, and suitable band-edge potential corresponding to redox potentials must be formed within the photocatalyst crystallography. Even then, the photo-generated charges must diffuse through a highly crystalline structure and interact with molecules and ions at highly reactive surface sites. Favorably, graphene as an abundant and cost-effective compound, is among the few nonmetallic photocatalysts that meet all the required conditions.

The molecular energy state of carbon is unique and contains $1s^2$, $2s^2$, and $2p^2$ orbitals, which contribute accordingly in various crystalline structures. As displayed in **Figure 2a**, the sp^2 hybridized structure is formed from s, p_x, and p_y orbitals on each carbon atom connected to the surrounding atoms through three strong covalent σ bonds. In this structure, the remaining $2p_z$ orbital, perpendicular to the graphene plane, overlapped with the one in the neighboring atoms, creating delocalized π (fill band) and π^* (empty band) orbitals, which are also known as graphenes VB and CB [35, 36].

The Fermi level in pristine graphene is located at the points connecting the valence and conduction bands in momentum space, as shown in **Figure 2b**, which is also known as the Dirac point [36]. Graphene exhibits an intrinsic n-type character [38]. Because of a small overlap between its valence and conduction bands, graphene has been characterized as a semimetal and/or a

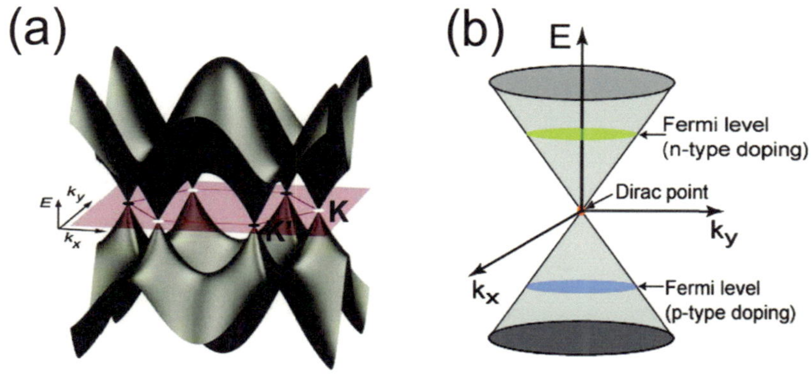

Figure 2. (a) The 3D band structure of graphene. (b) The linear dispersion and the band structure at the Dirac point in graphene [37]. Reprinted with permission from [37]. Copyright 2010 American Chemical Society.

semiconductor with zero band gap energy [13, 39, 40]. This orbital structure creates high conductivity greater than that for silver, which is the least resistive metallic material [41].

By negative and positive doping of graphene's lattice, the state of the Fermi level around the Dirac point can be tuned; therefore, the electronic characteristics of graphene can be tailored. This unique characteristic is attributed to the increase in the concentration of charge carriers by two orders of magnitude, to obtain n- or p-type graphene [17, 35]. For instance, the Fermi level of graphene can be shifted below the Dirac point so p-type characteristics can be formed between graphene and the rutile TiO_2 (110) surface, whereas the holes are accumulated on graphene and electrons are localized in the CB energy state of TiO_2 [38]. Therefore, the graphene electronic structure can be considered a photocatalyst design enabler, as the nanoscale p-n junction can be formed between localized p- and n-doped islands within the 2D crystallography.

From defect-free monolayer graphene to 3D aggregates of multilayer graphene, graphene nanoribbons, rGO, and finally GO, the band gap can be tuned within a wide range through engineering the morphology of the synthesized sample or surface modifications with functional groups. For instance, graphene nanoribbons are 1–100 nm wide, long strips of graphene, where a band gap is formed in the band structure of nanoribbon graphene when the width of the strips is reduced to less than 20 nm [42]. Due to such strange characteristics, graphene nanoribbons, nanomesh, and quantum dots exhibit semiconducting behavior with a band gap less than 0.5 eV [36]. Another strategy to engineer the band gap of graphene derivatives is intermolecular hybridization. An increase in the conductivity of graphene as a result of incorporation with another carbon compound was reported. For example, graphene derivatives–carbon nanotube composites, which are promising materials for transparent conductive materials, exhibited reduced surface resistance from their original 660 Ω sq.$^{-1}$ and 890 Ω sq.$^{-1}$ for rGO, and multiwalled CNTs, respectively, to ~100 Ω sq.$^{-1}$ for the hybrid, although hybridization reduces the optical transmittance of the composite [43]. Similar observations were also reported for CNTs grown on graphene utilized for optoelectronic applications [44–46].

Owing to its large band gap, GO is characterized as an insulating material. Optical measurements of the carriers' lifetime in various wavelengths show an ns-scale decay, indicating semiconducting behavior of GO [47]. Chemical, electrochemical, and photo-induced reduction of GO to rGO are facile and functional routes to tune the band gap of semiconducting nanosheet-like compounds over a wide range, with conductivity up to 30,000 Sm^{-1} [30, 47]. Mathkar et al. [48] studied the controlled reduction of GO using hydrazine vapor while monitoring the optical and compositional properties of the obtained rGO. They successfully controlled the band gap of rGO from 3.5 eV (sp^3 rich) to 1.0 eV (sp^2 rich), as demonstrated in **Figure 3a**.

A similar transition pattern was reported by Mattevi et al. [49] who investigated the surface composition and structure of GO at various stages of thermal reduction. As presented in **Figure 3b**, functionality was developed between the density of sp^2 bonds and the reduction-deriving force (reduction temperature) that exhibited the transformation of an individual GO nanosheet to rGO with a variety of electronic properties. Taking this transition pattern into

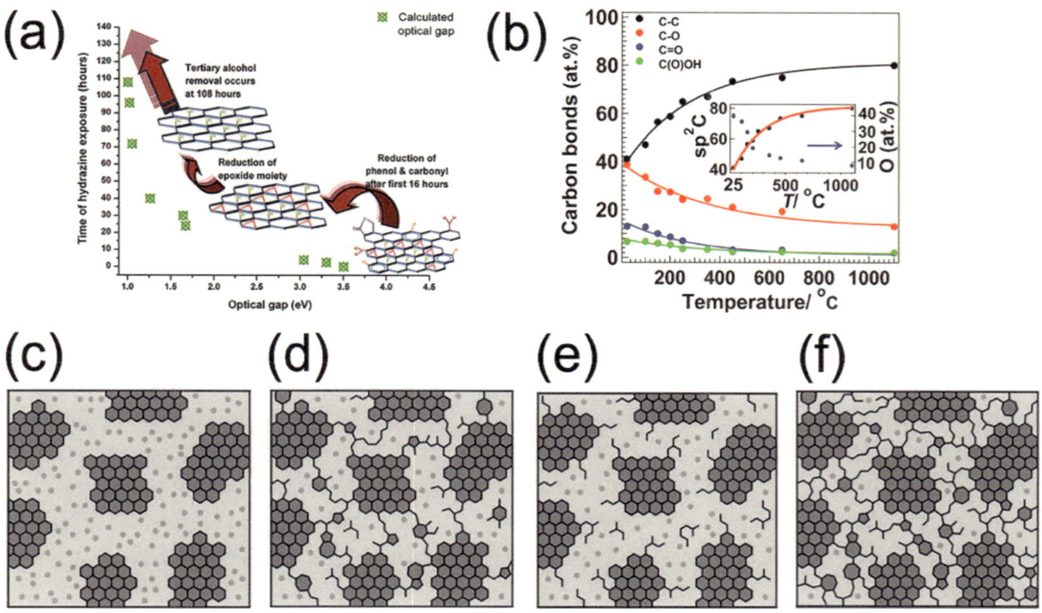

Figure 3. (a) Gradual chemical reduction of GO: the band gap energy and schematic structure of rGO at various stages of the chemical reduction showing E_g changes from 3.5 to 1.0 eV [48]. Reprinted with permission from [48]. Copyright 2012 American Chemical Society. (b) Percentage of different carbon bonds as a function of GO thermal reduction obtained via XPS. Inset: sp^2 carbon and the corresponding oxygen concentration [49]. (c–f) Structural model of GO at various stages of thermal reduction: Lattice at (c) room temperature; (d) ~100°C; (e) ~220°C; and (f) ~500°C, where the dark gray area indicates sp^2 clusters and the light gray area represents sp^3 bonds to oxygen-containing groups [49]. Reprinted with permission from [49]. Copyright 2009 Wiley-VCH Verlag GmbH & Co. KGaA, Weinheim.

consideration, the thermal energy was shown to be very effective in the reduction of GO and the preparation of conductive film at the early stages (T < 700°C); even at excessive temperatures (T~1000°C), the entire sp^3 network cannot be eliminated and complete healing of the C–C bonds cannot be realized. From this result, Mattevi and coworkers [49] proposed a mechanistic model to describe the evolving conductive sp^2 network. As shown in **Figure 3c–f**, in the course of thermal reduction, the connection between the original conductive graphitic clusters (shown as dark gray in **Figure 3c**) is formed, which leads to development of the sp^2 network (**Figure 3d**). Electrical conductivity of polycrystalline graphite (1.25×10^3 S cm^{-1}) can be obtained at an sp^2 fraction of ~0.87 when a great portion of the sp^2 islands are connected via an sp^2 network (**Figure 3e**) and further enhanced up to the point that approaches the one for single-layer graphene (6×10^3 S cm^{-1}) at an sp^2 fraction of 0.95 (**Figure 3f**).

To accomplish a suitable band gap for solar water splitting and CO_2 reduction reactions, the graphene sp^2 network must be extensively interrupted by functionalization of various oxygenated groups. Thus, the interaction between molecular orbitals of carbon atoms and oxygen functionalities forms a forbidden electronic band as high as 2.7 eV, while maintaining sufficient carrier mobility to transfer the photo-induced charges. Crystalline defects create preferential bonding sites for the adsorption and deposition of atoms and molecules, which can be employed for the fabrication of interaction with active species in electrolyte. However, more importantly, defects in the structure of graphene significantly enhance the density of dangling C—O bonds at the edges, which is very reactive for various redox reactions. The density of crystallographic defects must be optimized, as in contrast to aforementioned advantages; high density of structural defects increases the resistivity of graphene.

GO has exhibited photo-induced activity for reduction reactions by promoting electrons to their conduction band upon photon absorption. For example, GO was used for the UV-assisted reduction of biological samples (reduction of resazurin to resorufin), showing no sign of degradation at 350 nm [24]. Yeh et al. [50] investigated the functionality of GO for UV–Vis-induced sacrificial hydrogen evolution from 20% MeOH aqueous solution. Over the course of the reaction, in addition to the proton reduction at the active sites, evidently a portion of the photo-induced electrons interacted with the surface oxygen functional groups; thus, the band gap of the spent photocatalyst was reduced to 2.4 eV from its original 4.3 eV. Interestingly, the bare GO sample generated 17,000 μmol within 6 h under a 400 W high pressure mercury lamp, which was considered higher than the one loaded with 5 wt% Pt and far above the amount of hydrogen obtained from pure water (280 μmol) for the same period (**Figure 4a**). The low activity of the Pt-loaded sample is likely attributed to surface coverage of GO with opaque Pt nanoparticles.

GO has also exhibited visible-light ($\lambda > 400$ nm) activity in MeOH solution, as shown in **Figure 4b**, reached QE = 0.01%, which is significantly lower than that observed under a mercury lamp (QE = 2.70%). The stable evolution of hydrogen even at low band gap energy not only suggests that the CB of semiconducting GO is laid down at a suitable energy state corresponding to H^+/H_2 potential but also indicates that the transitional decay in the GO band gap is attributed to the upward shift of the VB, which is responsible for the lack of O_2 evolution, even in the presence of a scavenging Ag^+ ion [50]. This hypothesis can be extracted from the spectroscopic measurements, indicating that the removal of oxygen-containing groups on the surface leads to the reduction in the band gap through shifting the VB maximum upward, while the CB potential remained nearly unchanged at −0.75 to −0.71 eV versus Ag/AgCl [51, 52]. Further oxidization of GO increases the band gap and provides sufficient overpotential at the GO molecular orbital for an O_2 evolution reaction. However, the activity of the photocatalyst is expected to decline via photo-reduction, as shown for GO before and after photo reduction in **Figure 5a** and **b**, respectively [51].

The valence band maximum (VBM) and conduction band minimum (CBM) potential of GO can be tuned through doping, so within the localized sp^2 islands, both p- and n-type conductivities can be formed [30]. Therefore, individual islands may be activated for one photocatalytic half-reaction corresponding to their electrical characteristics. GO is intrinsically p-type due to electron-withdrawing oxygen functional groups on its surface [17]. Replacing these oxygen-containing groups with nitrogen, via noninvasive routes such as high temperature ammonolysis [53], results in n-type GO [30]. N-doped graphene oxide quantum dots (NGO–QDs) containing 6% carbon-bonding composition with N1 s/C1s = 2.9% posed a 2.2 eV band gap and exhibited overall water splitting under visible-light irradiation, comparable to that of the $Rh_{2-y}Cr_yO_3$ loaded GaN:ZnO solid solution photocatalyst, without the use of any precious metals [53]. The origin of this activity can be explained through the co-existence of p- and n-type conductivity within the lattice of the N-doped graphene QDs divided by the original sp[2] islands (ohmic contacts). Thus, the photo-induced electrons and holes recombined at this contact, providing charges at suitable energy states for overall water splitting, is demonstrated in **Figure 6**.

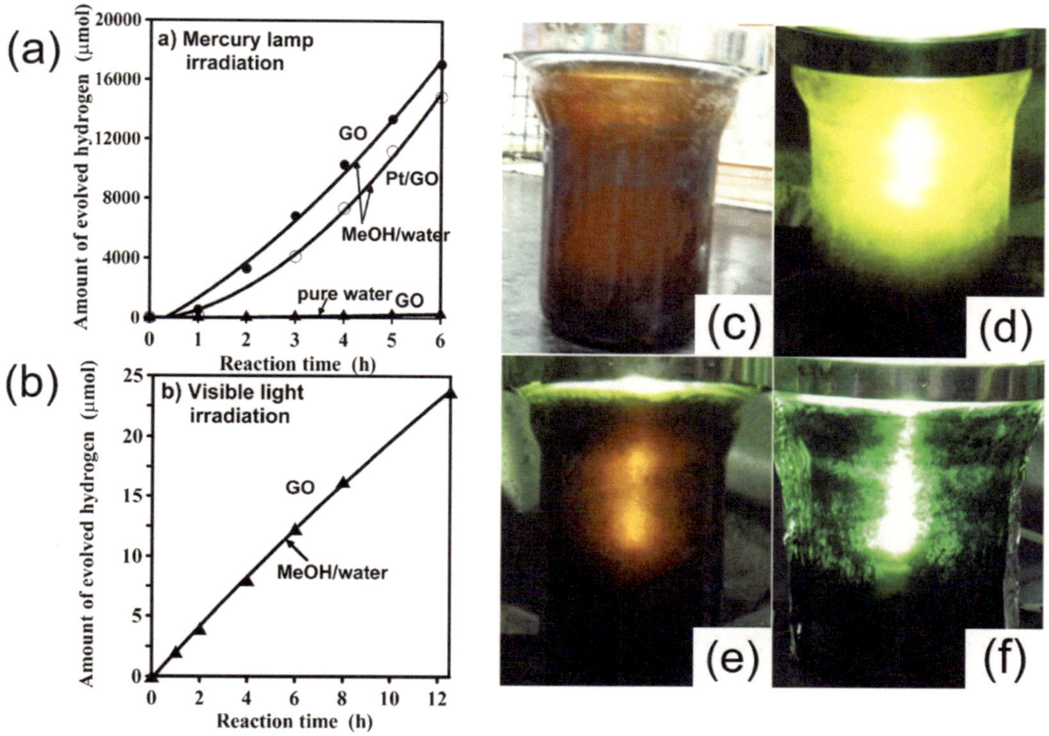

Figure 4. Hydrogen evolution from 20% MeOH solution and pure water using GO and 5 wt% Pt loaded GO photocatalysts (a) under UV–Vis irradiation and (b) under Vis irradiation. (c–f) Color variation of GO photocatalyst in a 20% MeOH solution indicating the self-reduction of GO (c) before irradiation; (d) at the start of irradiation; (e) after 30 min; and (f) after 2 h [50]. Reprinted with permission from [50]. Copyright 2010 Wiley-VCH Verlag GmbH & Co. KGaA, Weinheim.

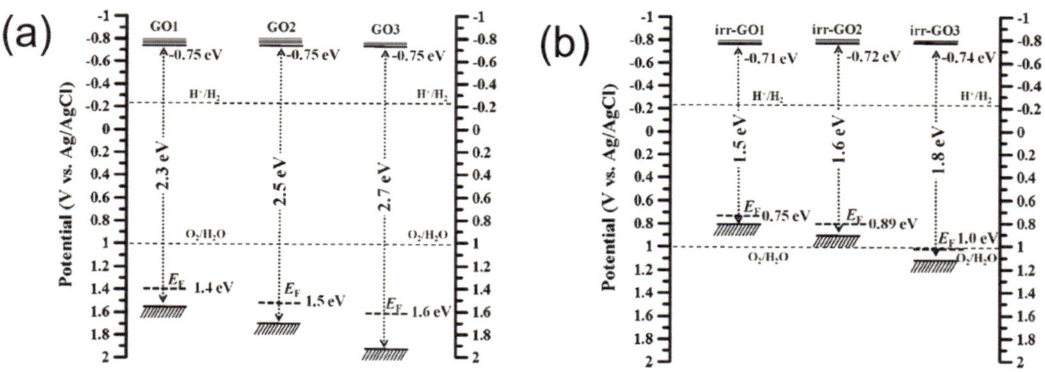

Figure 5. Band structure of GO samples (a) before and (b) after UV irradiation. GO1, GO2, and GO3 denoted for GO samples processed at oxidized through Hummers method for 4, 12, and 24 h [51]. Reprinted with permission from [51]. Copyright 2011 American Chemical Society.

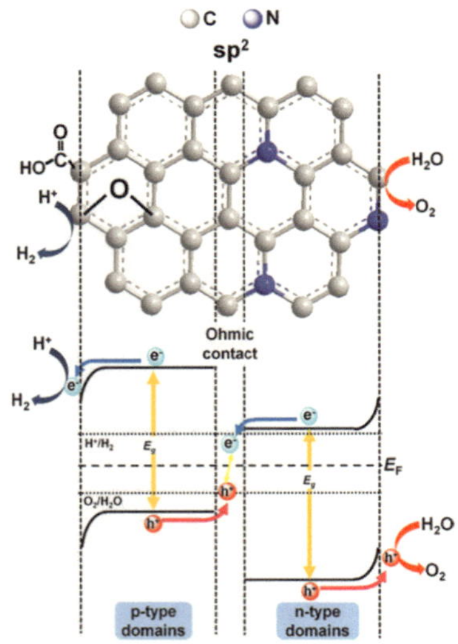

Figure 6. Mechanism of overall water splitting on p-type and n-type conductivity in GO lattice interconnected with sp^2 ohmic contact [53]. Reprinted with permission from [53]. Copyright 2014 Wiley-VCH Verlag GmbH & Co. KGaA, Weinheim.

Tan et al. [54], for the first time, reported the activity of isolated GO nanosheets for the conversion of CO_2 to methane (0.0628 μmol g^{-1} h^{-1} for 2 h) under visible light (15 W energy-saving daylight bulb source), although in situ photo-induced removal of oxygen functional groups resulted in a gradual decrease in methanol generation. CO_2 conversion to MeOH was also reported over modified graphene oxides with various band gap energies [55]. The phosphate-modified graphene with band gap ~3.2–4.4 eV produced methanol from CO_2 at 0.172 μmol gr^{-1} h^{-1}, nearly six times higher than the one measured for commercial TiO_2 under a 300 W halogen lamp. Since the photoexcitation in GO attributed to the surface oxygenated bonds, in this system, surface modification facilitated the electron/hole pair generation, which resulted in oxygen evolution and CO_2 reduction at the VB and CB, respectively [53].

Due to the high redox activity of crystalline defects in 2D carbon structures, graphene hybridized light absorber antenna have been the subject of photo-electro-chemical research. Dye sensitization has emerged as an effective technique to harvest a visible portion of the solar spectrum, and an alternative to the utilization of costly inorganic semiconductors [56–60]. A positive charge on the surface of dye molecules promotes their association and interfacial contact with negatively charged GO through electrostatic attraction. Latorre-Sánchez and coworkers [52] have functionalized the surface of GO with various degrees of oxidation (~10% carbon-oxygen content), hybridized it with a series of dye molecules (up to 39 wt%), and tested the fabricated composite for H_2 evolution in 20 V% methanol solution under 532 nm monochromatic light. In their experiment, the composite of cationic and anodic Ru dye complexes were anchored to the interlayer space of multilayer GO. The cationic $[Ru(bipy)_3]^{2+}$ exhibited the highest total corrected hydrogen evolution (with regard to dye absorption spectra), while the anodic Ru polypyridyl complex (N719) demonstrated the highest hydrogen evolution rate after nearly 100 min induction period. The induction period for N719 dye is likely attributed to the activation period, and/or the relocation

http://dx.doi.org/10.5772/intechopen.72623

of dye molecules through the redox reaction. Pt-loaded rGO nanosheets as electron mediators for Eosin Y (EY) dye molecules has exhibited 9.3% QE under visible light, which is significantly greater than those recorded for EY-Pt and EY-rGO composites (**Figure 7a**). Due to the position of the energy bands, the excited electrons are transferred from the CB of the dye molecule to rGO and eventually to Pt active sires, as demonstrated in **Figure 7b**.

Further tailoring the band structure is shown to be highly effective, as the 5 wt% Pt-loaded GO nanosheets cosensitized by 1:1 M EY: Rose Bengal (RB) dye molecules reached quantum yield as high as 37.3% in 15 V% TEOA solution under two 450 W Xe lamps adjusted for 520 and 550 nm irradiation [62].

3.2. Graphene as electron acceptor

The process of transportation of an excited electron-hole pair from an emitter or "donor" to an absorbing medium, or "acceptor" is called nonradiative energy transfer (NRET). According to a study by Raja and coworkers, the rate of NRET in layered materials such as graphene and MoS_2 is a function of number of layers by comparing the decay rates of quantum dot fluorescence when the chromophores are placed on graphene and MoS_2 [63]. As illustrated with a gray line in **Figure 8a**, the population of charge carriers decays relatively slowly (a luminescence lifetime of 5 ns) in the absence of the acceptor parties (graphene and MoS_2). After planting QDs on 2D materials, due to high NRET rate, the photoluminescence (PL) lifetime decay decreased by an order of magnitude. Interestingly, their study indicates that the PL lifetime (inversely proportional to NRET rate) drops and increases by increasing the number of layers in graphene and MoS_2, respectively (**Figure 8b**). This finding confirms the strong electron acceptor role of graphene and provides hints for fabrication of highly efficient energy conversion/storage and optoelectronic devices.

In the first demonstration of graphene-based electron acceptors, Liu et al. [64] reported a solution-processable functionalized graphene electron-accepter with poly(3-octylthiophene) and poly(3-hexylthiophene) donor materials, reaching power conversion efficiency of 1.4% at 100 mW cm^{-2} AM 1.5G. In 2008, Williams et al. [65] discovered a room-temperature technique for the in situ reduction of GO via the transfer of electrons from semiconductor bulk to nanosheets.

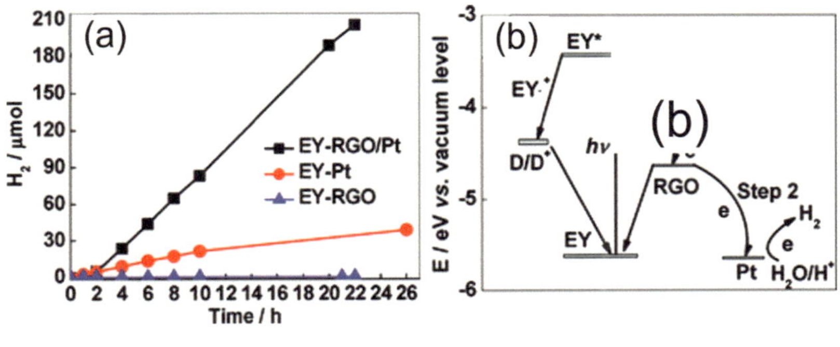

Figure 7. (a) The time courses of H_2 evolution over EY-Pt, EY-RGO, and EY-RGO-Pt photocatalysts and (b) photogenerated electron transfer from EY to rGO, and eventually to Pt, due to the composite energy level diagram [61]. Reprinted with permission from [61]. Copyright 2011 American Chemical Society.

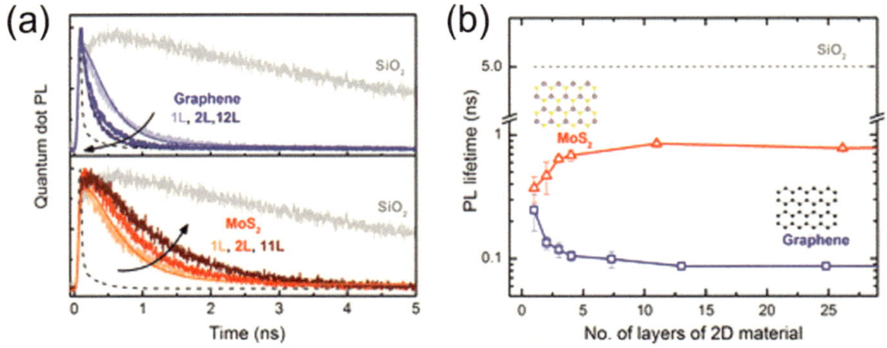

Figure 8. PL lifetime decay of QDs deposited on 2D graphene and MoS_2 (a) and the impact of graphene and MoS_2 thickness on the PL lifetime of charge carriers (b) [63]. Reprinted with permission from [63]. Copyright 2016 American Chemical Society.

This discovery facilitated the utilization of graphene derivatives as electron acceptors for solar fuel generation. The interesting work of Ng and coworkers [66] is one of the early studies of such efforts through the fabrication of $BiVO_4$ hybridized GO followed by in situ photo deposition.

In the last decade, various combinations of semiconductors and graphene-derivative electron acceptors and active sites were studied for effective solar fuel generation. Thus far, the combination of graphene derivatives and promising semiconducting photocatalyst via facile techniques was reported, including metal oxides [67–70], nitrides [71, 72], oxynitrides [26, 73], sulfides, and oxysulfides [74–77], as well as ternary composites [78] and those based on abundant Earth materials [79–81]. To realize the effectiveness of such hybridization, a recent study by Tang and coworkers [82] can be discussed. An advanced ternary photocatalytic system of TiO_2 nanotube-decorated CdS nanoparticles in composite with rGO nanosheets were synthesized and tested for photocatalytic hydrogen evolution where (1) the light absorption had been extended to the visible region; (2) the photo-excited charge separation significantly boosted and the electron path was configured to $CdS_{(CB)} \rightarrow TiO_{2(CB)} \rightarrow rGO$; and (3) CdS photo-corrosion was suppressed because of the rGO nanosheet protection (60 h stable H_2 evolution), via the so-called sheltering effect.

Readers are encouraged to read interesting review articles published recently on related topics, such as the one published by Xiang et al. [3] and Low et al. [6].

3.3. Graphene as electron mediator

Ternary nanocomposites of semiconductors and carbon-based materials can be designed and fabricated to tailor the absorption spectrum and to enhance the catalytic performance for solar fuel generation. The visible-light-responsive semiconductors were incorporated into the existing composites to further absorb the incident photons [82–85]. Effective electronic interaction between two photo-responsive components forms a recombination contact between the two and thus suppresses the unwanted recombination lost within one intrinsic component.

Hou et al. [78] reported an outstanding visible-light hydrogen evolution over the CdS-core TaON-shell rGO ternary composite prepared via the hydrothermal-assisted ion-exchange technique. In this composite, rGO serves as an active site for H_2 evolution, as discussed in Section 3.1, as well as an electron mediator for efficient charge transfer within composite components. Due to the higher VB level of CdS, the photo-excited holes are promptly transferred to CdS and eventually to electrolyte and reduce the rate of recombination. By adding rGO and the Pt cocatalyst, the photo-induced electrons sink into rGO and subsequently into Pt, driven by their working function differentiation, which further reduces the recombination losses. The 1 wt% CdS-content ternary rGO hybridized core-shell composite decorated with 0.4% Pt reached 31% sacrificial apparent quantum efficiency at 420 nm (Na_2S-Na_2SO_3 aqueous solution), over twice that of the one without rGO and over 140 times of the bare TaON.

Z-scheme photocatalytic systems have demonstrated tremendous potential for efficient solar energy conversion. The first report on the stoichiometric water splitting into H_2 and O_2 through the Z-scheme mechanism was published in 1997 by Sayama et al. [86]. Since then, extensive studies have placed emphasis on the structure and catalytic behavior of individual Z-schematic components and their electronic interaction. Z-scheme water splitting is particularly of interest because the wide range of reduction and oxidation light-absorbing components provides design flexibilities for solar fuel generation. The conventional Z-scheme design is adapted through electron transfer from an O_2-evolving photocatalyst to an H_2-evolving photocatalyst via ionic electron mediators, such as IO^{3-}/I^- and Fe^{3+}/Fe^2 [87, 88]. Since the electron mediators must efficiently transfer electrons by adsorbing and desorbing onto and from the surfaces of photocatalysts, unstable semiconductors, such as metal sulfides, are not considered good candidates for Z-scheme water splitting [89]. Besides its role as an electron acceptor in semiconductor composites, graphene derivatives were used as a solid-state electron mediator for Z-scheme photocatalytic systems. Graphene-based compounds are excellent conductive platforms as solid-state electron mediators and provide enormous possibilities for the commercialization of Z-scheme solar-fuel-generation technology.

Recently, Iwashina et al. [89] synthesized a series of p-type metal sulfides (HER) in ternary composite with n-type rGO-TiO_2(OER). Among the various composites, $CuGaS_2$ loaded with 0.1 wt% Pt cocatalyst exhibited the highest activity for water splitting, yielding QE = 1.3% under 380 nm monochromatic irradiation. In this system, no appreciable gas evolution was observed in the absence of rGO, and/or either Pt-loaded $CuGaS_2$ or TiO_2, indicating the Z-schematic mechanism and the efficient contribution of rGO as an electron mediator.

The effectiveness of rGO as a solid-state electron mediator for Z-schematic water splitting consists of the $BiVO_4$ as O_2-evolution photocatalyst and Rh-doped $SrTiO_3$ decorated with Ru-complex cocatalyst $[Ru(2,2'\text{-bipyridine})(4,4'\text{-diphosphonate-}2,2'\text{-bipyridine})(CO)_2]^{2+}$ as H_2-evolving photocatalyst was investigated, as demonstrated in **Figure 9a**, which suggests enormous potential for electrolyte-independent overall water splitting [90, 91]. The electrons in the CB of $BiVO_4$ and the holes in the VB of $SrTiO_3$:Rh cannot meet the energy requirements for water reduction and oxidization, respectively. Therefore, as depicted in **Figure 9b**, upon visible light excitation, the electrons in the CB of $BiVO_4$ and the holes in the electron-donor level of $SrTiO_3$:Rh are transferred to the rGO mediator and recombined at the conductive support. The free electrons and holes on the $SrTiO_3$:Rh and $BiVO_4$ surface freely participated in the H_2 and O_2 evolution, respectively. Due to this effective electron mediator characteristic of rGO, as shown in **Figure 9c**, the stoichiometric and stable visible-light driven Z-scheme overall water splitting for the $BiVO_4$-rGO-$SrTiO_3$:Rh decorated with Ru-based HER cocatalyst system was observed.

The complex catalysts based on ruthenium have been extensively studied due to their potential in mimicking the plant photosynthesis process [92]. In particular, Ru complexes demonstrated stable and promising performances for CO_2 conversion with high selectivity toward HCOOH [93]. The ruthenium complex reduction catalyst containing 2-thiophenyl benzimidazole ligands (schematic **Figure 10**) was studied for visible-light CO_2 reduction while it was covalently anchored to GO through the epoxide groups on the GO surface [94]. The treatment of GO with chloroacetic acid leads to conversion of the —OH and epoxide groups to —COOH groups, which was further treated by thionyl chloride to transform GO—COOH to —COCl functionalized GO. This 2D platform is ideal for interaction with the Ru-complex, as schematically illustrated in **Figure 10**. The Ru-complex-GO composite produced over 2000 μmol g^{-1} of formic acid in 20 h, without use of a sacrificial reagent.

Kuai and coworkers [95] fabricated TiO_2–CdS encapsulated rGO composite where graphene was used as a solid-sate electron mediator for the Z-schematic conversion of CO_2 in the presence of water vapor. Hydrothermally prepared CdS nanospheres, as shown in **Figure 11a** and **b**, were positively charged and wrapped inside rGO nanosheets via self-assembly, which was induced by electrostatic forces (**Figure 11c** and **d**), followed by kinetic-controlled deposition of TiO_2 nanoparticles, as illustrated in **Figure 11e**. Such nanostructure is highly beneficial for Z-scheme photocatalysis, since rGO is positioned between two redox antenna/active sites, as demonstrated in **Figure 11f**, where ideal interfacial contact between the composite components can be maintained. Thus, the photo-induced electrons in TiO_2 CB are transferred to rGO and subsequently recombine with the holes generated at the CdS nanosphere's VB, resulting in enhanced density of photo-generated electrons and holes on the CdS nanosphere and TiO_2 nanoparticles, respectively. The photocatalytic efficiency of the CdS-rGO-TiO_2 Z-scheme system under 300 W Xe lamp irradiation was remarkably enhanced relative to pristine CdS, CdS-TiO_2, and CdS-rGO composites.

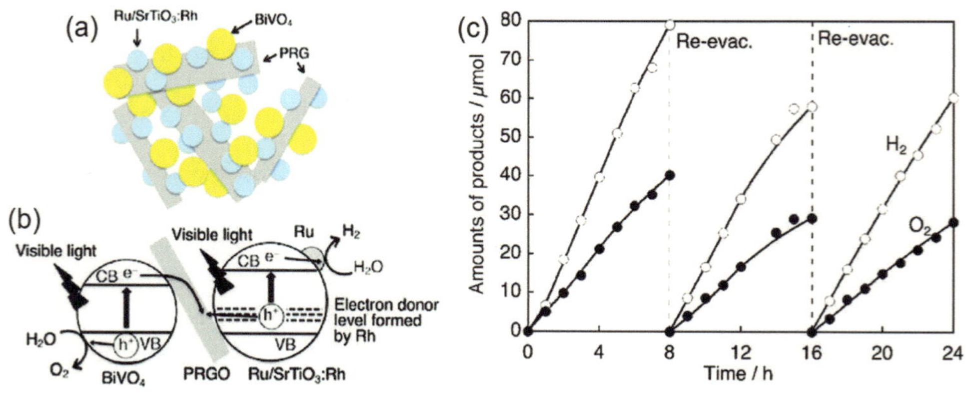

Figure 9. (a) Schematic illustration of rGO-mediated Z-scheme water splitting in Ru-SrTiO3:Rh (HER) and $BiVO_4$ (OER) system. (b) Photo-induced charge transfer driven by the energy band structure of the composite. (c) Time-course of visible light Z-scheme overall water splitting of Ru-$SrTiO_3$:Rh-rGO-$BiVO_4$ under a 300 W Xe lamp [90]. Reprinted with permission from [90]. Copyright 2011 American Chemical Society.

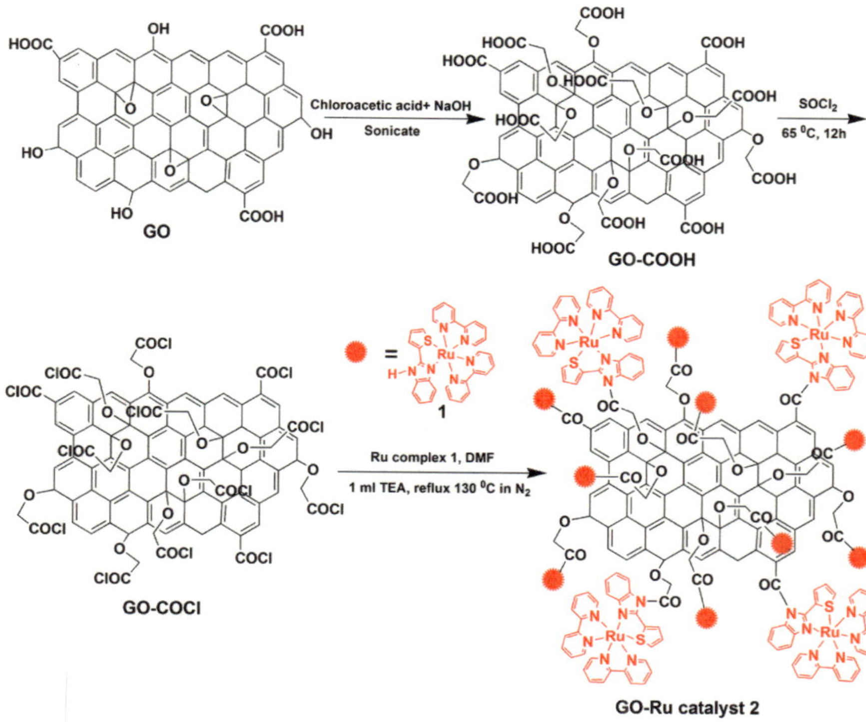

Figure 10. Surface functionalization of GO with a Ru catalyst complex [94]. Reprinted with permission from [94]. Copyright 2015 Royal Society of Chemistry.

3.4. Graphene as thermally-induced medium

Until now, the majority of studies on graphene-based photocatalysts have placed their focus on the role of graphene derivatives as transparent/flexible electrodes, electron acceptors, and electron mediators. Recently, the functionality of graphene-based compounds was investigated as a light absorber. In the recent years, it was concluded that graphene could absorb the entire solar spectrum because of its black color and zero band gap. Although such capability does not lead to active electron/hole pair generation for photocatalytic reaction, it does result in local high-temperature zones inside and on the surface of the photocatalyst, due to the photo-thermal effect, and enhances photocatalytic activity. According to a study by Gan and coworkers [96], such a photo-thermal effect contributes up to 38% in the photo-degradation of organic pollutants in the P25-rGO composite.

Colloidal QDs exhibit high quantum efficiencies through band gap engineering methods and are known to be excellent absorbers and emitters at various wavelengths [63]. Graphene QDs can be prepared via green and facile techniques [97], and have demonstrated strong quantum confinement at sizes below 10 nm [98]. The challenge of employing the interesting PL properties of graphene QDs for visible-light photocatalytic applications is their excitation-dependent PL properties, which means that at various excitation wavelengths, different energies are induced in the orbital structure of QDs. Zhuo et al. [98] prepared graphene QDs through the facile ultrasound route and exhibited an excitation-independent PL peak at 407 nm. Due to this constant PL emission, rutile

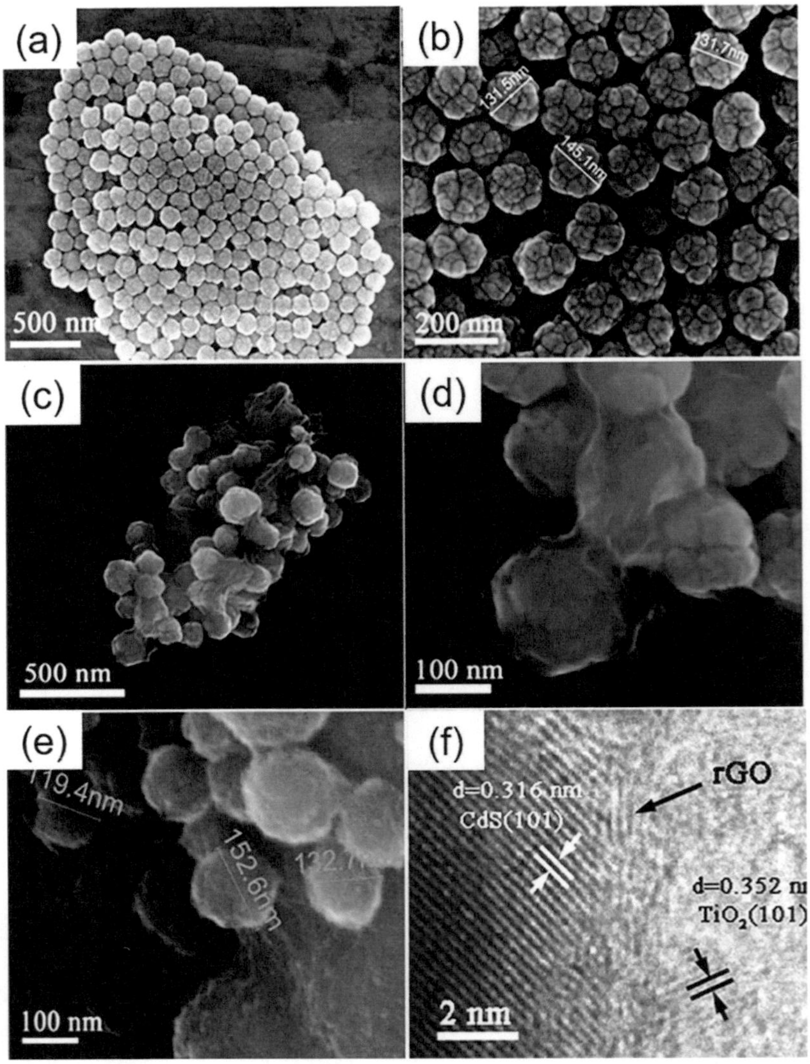

Figure 11. SEM images of CdS NSs (a and b), CdS NSs/GO (c and d), CdS NSs/rGO/TiO_2 (e) and the HRTEM image of CdS NSs/rGO/TiO_2 (f) [95]. Reprinted with permission from [95]. Copyright 2015 Royal Society of Chemistry.

TiO_2-QDs showed nine-fold visible-light catalytic activity (since it has a narrower band gap than 407 nm), compared to that of anatase TiO_2 (a band gap larger than 407 nm).

Recently, free-standing vertical graphene exhibited high potential as a light absorber. Zhao and coworkers [99] proposed a detailed mechanism for plasma-enhanced chemical vapor deposition of free-standing vertical graphene nanosheets, which is validated by transmission electron microscopy observations at the nucleation and growth stages. Davami et al. [100] reported that such morphology absorbs up to 97% of incident photons, which is much higher than previous carbon-based absorber materials, such as forest CNTs. Although free-standing graphene absorbers are yet to be studied for the solar-fuel-generation application, preliminary research data suggest their great potential for future-generation photocatalytic applications.

4. Conclusion and perspective

Graphene-derivatives, such as graphene oxide, reduced graphene oxide, and their functionalized materials are attractive components in optoelectronic device fabrication, owing to their peculiar optical, thermal, mechanical, and electrochemical properties. In particular, the 2D carbon nanostructure has proven to be an excellent candidate for the extraction of solar energy and its chemical transformation to fuel through photocatalytic water splitting and CO_2 conversion. Graphene derivatives' ultra-high specific surface area, which promotes the rate of redox reactions; its tunable band gap, which controls their absorption spectrum; its controllable defects density, which promotes their reactivity; and its carrier mobility, which promotes charge separation, all offer design flexibility at the nanoscale. However, the performance of graphene-based photocatalytic systems is vastly unpredictable, and understanding the role of graphene in photocatalytic fuel generation is still under debate. Over 18,000 articles related to graphene were published from 2004 to 2014. This rapid progress seeks continuous publication of critical reviews to genuinely add to the existing literature and to discuss the state-of-the-art development. Looking at the fast-paced advances in nanotechnology and materials sciences, it is apparent that the use of solar energy is an indispensable reality. The sunlight-induced splitting of waste water and atmospheric CO_2 reduction offers the onsite production of clean and renewable energy, as well as bacterial disinfection and pollutant decomposition. Therefore, the search for a highly active photocatalyst that can be produced through low-cost and scalable routes, exhibits a stable performance, and does not pose any threat to the environment is an extremely important task. Graphene derivatives are among the few candidates that meet all the aforementioned conditions.

Author details

Babak Adeli[1] and Fariborz Taghipour[1,2*]

*Address all correspondence to: fariborz.taghipour@ubc.ca

1 Photoreaction Engineering Laboratory, Chemical and Biological Engineering Department, University of British Columbia, Vancouver, BC, Canada

2 Clean Energy Research Centre (CERC), University of British Columbia, Vancouver, BC, Canada

References

[1] Adeli B, Taghipour FA. Review of synthesis techniques for gallium-zinc oxynitride solar-activated photocatalyst for water splitting. ECS Journal of Solid State Science and Technology. 2013;**2**:Q118-Q126

[2] Crabtree GW, Lewis NS. Solar energy conversion. Physics Today. 2007;**60**:37-42

[3] Xiang Q, Cheng B, Yu J. Graphene-based photocatalysts for solar-fuel generation. Angewandte Chemie International Edition. 2015;**54**:11350-11366

[4] Adeli B, Taghipour F. Facile synthesis of highly efficient nano-structured gallium zinc oxynitride solid solution photocatalyst for visible-light overall water splitting. Applied Catalysis A: General. 2016;**521**:250-258

[5] Huang X, Yin Z, Wu S, et al. Graphene-based materials: Synthesis, characterization, properties, and applications. Small. 2011;**7**:1876-1902

[6] Low J, Yu J, Ho W. Graphene-based Photocatalysts for CO2 reduction to solar fuel. Journal of Physical Chemistry Letters. 2015;**6**:4244-4251

[7] An X, Yu JC. Graphene-based photocatalytic composites. RSC Advances. 2011;**1**:1426

[8] Xiang Q, Yu J. Graphene-based photocatalysts for hydrogen generation. Journal of Physical Chemistry Letters. 2013;**4**:753-759

[9] Sun Y, Wu Q, Shi G. Graphene based new energy materials. Energy & Environmental Science. 2011;**4**:1113

[10] Novoselov KSS, Geim AKK, Morozov SVV, et al. Electric field effect in atomically thin carbon films. Science (80-). 2004;**306**:666-669

[11] Inagaki M, Kim Y a., Endo M. Graphene: Preparation and structural perfection. Journal of Materials Chemistry 2011;**21**:3280

[12] Hazra KS, Sion N, Yadav A, et al. Vertically aligned graphene based non-cryogenic bolometer. 2013. http://arxiv.org/abs/1301.1302 [Accessed: 12 July 2017]

[13] Garg R, Dutta NK, Choudhury NR. Work function engineering of graphene. Nanomaterials. 2014;**4**:267-300

[14] Miró P, Audiffred M, Heine T, et al. An atlas of two-dimensional materials. Chemical Society Reviews. 2014;**43**:6537-6554

[15] Huang X, Qi X, Boey F, et al. Graphene-based composites. Chemical Society Reviews. 2012;**41**:666-686

[16] Xiang Q, Yu J, Jaroniec M. Graphene-based semiconductor photocatalysts. Chemical Society Reviews. 2012;**41**:782

[17] Loh KP, Bao Q, Ang PK, et al. The chemistry of graphene. Journal of Materials Chemistry. 2010;**20**:2277

[18] Nair RR, Blake P, Grigorenko AN, et al. Fine structure constant defines visual transperency of graphene. Science (80-). 2008;**320**:2008

[19] Huang X, Zeng Z, Fan Z, et al. Graphene-based electrodes. Advanced Materials. 2012; **24**:5979-6004

[20] Chabot V, Higgins D, Yu A, et al. A review of graphene and graphene oxide sponge: Material synthesis and applications to energy and the environment. Energy & Environmental Science. 2014;**7**:1564

[21] Hsu HC, Shown I, Wei HY, et al. Graphene oxide as a promising photocatalyst for CO_2 to methanol conversion. Nanoscale. 2012;**5**:262-268

[22] Yeh TF, Cihlar J, Chang CY, et al. Roles of graphene oxide in photocatalytic water splitting. Materials Today. 2013;**16**:78-84

[23] Compton OC, Nguyen ST. Graphene oxide, highly reduced graphene oxide, and graphene: Versatile building blocks for carbon-based materials. Small. 2010;**6**:711-723

[24] Krishnamoorthy K, Mohan R, Kim S-J. Graphene oxide as a photocatalytic material. Applied Physics Letters. 2011;**98**:24-27

[25] Chen D, Feng H, Li J. Graphene oxide: Preparation, functionalization, and electrochemical applications. Chemical Reviews. 2012;**112**:6027-6053

[26] Adeli B, Taghipour F. Reduced graphene oxide composite of gallium zinc oxynitride photocatalyst with improved activity for overall water splitting. Chemical Engineering and Technology. 2016;**39**:142-148

[27] Ito J, Nakamura J, Natori A. Semiconducting nature of the oxygen-adsorbed graphene sheet. Journal of Applied Physics. 2008;**103**:113712

[28] Gao W. The chemistry of graphene oxide. Graphene Oxide: Reduction Recipes, Spectroscopy, and Applications. 2015;**39**:61-95

[29] Johari P, Shenoy VB. Modulating optical properties of graphene oxide: Role of prominent functional groups. ACS Nano. 2011;**5**:7640-7647

[30] Loh KP, Bao Q, Eda G, et al. Graphene oxide as a chemically tunable platform for optical applications. Nature Chemistry. 2010;**2**:1015-1024

[31] Pinto H, Jones R, Goss JP, et al. Unexpected change in the electronic properties of the au-graphene interface caused by toluene. Physical Review B: Condensed Matter and Materials Physics. 2010;**82**:1-8

[32] Yu S, Zhong Y-Q, B-Q Y, et al. Graphene quantum dots to enhance the photocatalytic hydrogen evolution efficiency of anatase TiO_2 with exposed {001} facet. Physical Chemistry Chemical Physics. 2016;**18**:20338-20344

[33] Yang KD, Ha Y, Sim U, et al. Graphene quantum sheet catalyzed silicon photocathode for selective CO_2 conversion to CO. Advanced Functional Materials. 2016;**26**:233-242

[34] Li L, Wu G, Yang G, et al. Focusing on luminescent graphene quantum dots: Current status and future perspectives. Nanoscale. 2013;**5**:4015

[35] Liu H, Liu Y, Zhu D. Chemical doping of graphene. Journal of Materials Chemistry. 2011;**21**:3335

[36] Lu G, Yu K, Wen Z, et al. Semiconducting graphene: Converting graphene from semimetal to semiconductor. Nanoscale. 2013;**5**:1353

[37] Avouris P. Graphene: Electronic and photonic properties and devices. Nano Letters. 2010;**10**:4285-4294

[38] Du A, Ng YH, Bell NJ, et al. Hybrid graphene/titania nanocomposite: Interface charge transfer, hole doping, and sensitization for visible light response. Journal of Physical Chemistry Letters. 2011;**2**:894-899

[39] Berger C, Song Z, Li X, et al. Electronic confinement and coherence in patterned epitaxial graphene. Science (80-). 2006;**312**:1191-1196

[40] Ando T. The electronic properties of graphene and carbon nanotubes. NPG Asia Materials. 2009;**1**:17-21

[41] Zhang X, Rajaraman BRS, Liu H, et al. Graphene's potential in materials science and engineering. RSC Advances. 2014;**4**:28987-29011

[42] Vicarelli L, Heerema SJ, Dekker C, et al. Controlling defects in graphene for optimizing the electrical properties of graphene nanodevices. ACS Nano. 2015;**9**:3428-3435

[43] Kim KS, Rhee KY, Park SJ. Influence of multi-walled carbon nanotubes on electrochemical performance of transparent graphene electrodes. Materials Research Bulletin. 2011;**46**:1301-1306

[44] Fan Z, Yan J, Zhi L, et al. A three-dimensional carbon nanotube/graphene sandwich and its application as electrode in supercapacitors. Advanced Materials. 2010;**22**:3723-3728

[45] Yu K, Lu G, Bo Z, et al. Carbon nanotube with chemically bonded graphene leaves for electronic and optoelectronic applications. Journal of Physical Chemistry Letters. 2011;**2**:1556-1562

[46] Velten J, Mozer AJ, Li D, et al. Carbon nanotube/graphene nanocomposite as efficient counter electrodes in dye-sensitized solar cells. Nanotechnology. 2012;**23**:85201

[47] Vempati S, Uyar T. Fluorescence from graphene oxide and the influence of ionic, π–π interactions and heterointerfaces: Electron or energy transfer dynamics. Physical Chemistry Chemical Physics. 2014;**16**:21183-21203

[48] Mathkar A, Tozier D, Cox P, et al. Controlled, stepwise reduction and band gap manipulation of graphene oxide. Journal of Physical Chemistry Letters. 2012;**3**:986-991

[49] Mattevi C, Eda G, Agnoli S, et al. Evolution of electrical, chemical, and structural properties of transparent and conducting chemically derived craphene thin films. Advanced Functional Materials. 2009;**19**:2577-2583

[50] Yeh T-F, Syu J-M, Cheng C, et al. Graphite oxide as a photocatalyst for hydrogen production from water. Advanced Functional Materials. 2010;**20**:2255-2262

[51] Yeh TF, Chan FF, Hsieh C, Te, et al. Graphite oxide with different oxygenated levels for hydrogen and oxygen production from water under illumination: The band positions of graphite oxide. Journal of Physical Chemistry C. 2011;**115**:22587-22597

[52] Latorre-Sánchez M, Lavorato C, Puche M, et al. Visible-light photocatalytic hydrogen generation by using dye-sensitized graphene oxide as a photocatalyst. Chemistry: A European Journal. 2012;**18**:16774-16783

[53] Yeh TF, Teng CY, Chen SJ, et al. Nitrogen-doped graphene oxide quantum dots as photocatalysts for overall water-splitting under visible light illumination. Advanced Materials. 2014;**26**:3297-3303

[54] Tan L-L, Ong W-J, Chai S-P, et al. Reduced graphene oxide-TiO_2 nanocomposite as a promising visible-light-active photocatalyst for the conversion of carbon dioxide. Nanoscale Research Letters. 2013;**8**:465

[55] Hsu H-C, Shown I, Wei H-Y, et al. Graphene oxide as a promising photocatalyst for CO_2 to methanol conversion. Nanoscale. 2013;**5**:262-268

[56] Choi SK, Yang HS, Kim JH, et al. Organic dye-sensitized TiO_2 as a versatile photocatalyst for solar hydrogen and environmental remediation. Applied Catalysis B: Environmental. 2012;**121-122**:206-213

[57] Choi SK, Kim S, Ryu J, et al. Titania nanofibers as a photo-antenna for dye-sensitized solar hydrogen. Photochemical & Photobiological Sciences. 2012;**11**:1437

[58] Tong L, Iwase A, Nattestad A, et al. Sustained solar hydrogen generation using a dye-sensitised NiO photocathode/$BiVO_4$ tandem photo-electrochemical device. Energy & Environmental Science. 2012;**5**:9472

[59] Gao Y, Ding X, Liu J, et al. Visible light driven water splitting in a molecular device with unprecedentedly high photocurrent density. Journal of the American Chemical Society. 2013;**135**:4219-4222

[60] Ding X, Gao Y, Zhang L, et al. Visible light-driven water splitting in photoelectrochemical cells with supramolecular catalysts on photoanodes. ACS Catalysis. 2014;**4**:2347-2350

[61] Min S, Lu G. Dye-sensitized reduced graphene oxide photocatalysts for highly efficient visible-light-driven water reduction. Journal of Physical Chemistry C. 2011; **115**:13938-13945

[62] Min S, Lu G. Dye-cosensitized graphene/Pt photocatalyst for high efficient visible light hydrogen evolution. International Journal of Hydrogen Energy. 2012;**37**:10564-10574

[63] Raja A, Montoya-Castillo A, Zultak J, et al. Energy transfer from quantum dots to graphene and MoS_2 : The role of absorption and screening in two-dimensional materials. Nano Letters. 2016;**16**:2328-2333

[64] Liu Z, Liu Q, Huang Y, et al. Organic photovoltaic devices based on a novel acceptor material: Graphene. Advanced Materials. 2008;**20**:3924-3930

[65] Williams G, Seger B, Kamt PV. TiO_2-graphene nanocomposites. UV-assisted photocatalytic reduction of graphene oxide. ACS Nano. 2008;**2**:1487-1491

[66] Ng YH, Iwase A, Kudo A, et al. Reducing graphene oxide on a visible-light $BiVO_4$ photocatalyst for an enhanced photoelectrochemical water splitting. Journal of Physical Chemistry Letters. 2010;**1**:2607-2612

[67] Xiang Q, Yu J, Jaroniec M. Enhanced photocatalytic H_2-production activity of graphene-modified titania nanosheets. Nanoscale. 2011;**3**:3670-3678

[68] Kim H, Moon G, Monllor-Satoca D, et al. Solar photoconversion using graphene/TiO_2 composites: Nanographene shell on TiO_2 core versus TiO_2 nanoparticles on graphene sheet. Journal of Physical Chemistry C. 2012;**116**:1535-1543

[69] Wang Y, Yu J, Xiao W, et al. Microwave-assisted hydrothermal synthesis of graphene based Au–TiO_2 photocatalysts for efficient visible-light hydrogen production. Journal of Materials Chemistry A. 2014;**2**:3847

[70] Zang Y, Li L, Zuo Y, et al. Facile synthesis of composite g-C_3N_4/WO_3: A nontoxic photocatalyst with excellent catalytic activity under visible light. RSC Advances. 2013;**3**:13646

[71] Ong W-J, Tan L-L, Chai S-P, et al. Surface charge modification via protonation of graphitic carbon nitride (g-C_3N_4) for electrostatic self-assembly construction of 2D/2D reduced graphene oxide (rGO)/g-C_3N_4 nanostructures toward enhanced photocatalytic reduction of carbon dioxide to methane. Nano Energy. 2015;**13**:757-770

[72] Xiang Q, Yu J, Jaroniec M. Preparation and enhanced visible-light photocatalytic H_2-production activity of graphene/C_3N_4 composites. Journal of Physical Chemistry C. 2011;**115**:7355-7363

[73] Mukherji A, Seger B, GQ L, et al. Nitrogen doped $Sr_2Ta_2O_7$ coupled with graphene sheets as photocatalysts for increased photocatalytic hydrogen production. ACS Nano. 2011;**5**:3483-3492

[74] Li Q, Guo B, Yu J, et al. Highly efficient visible-light-driven photocatalytic hydrogen production of CdS-cluster-decorated graphene nanosheets. Journal of the American Chemical Society. 2011;**133**:10878-10884

[75] Jia L, Wang DH, Huang YX, et al. Highly durable N-doped graphene/CdS nanocomposites with enhanced photocatalytic hydrogen evolution from water under visible light irradiation. Journal of Physical Chemistry C. 2011;**115**:11466-11473

[76] Xu J, Wang L, Cao X. Polymer supported graphene-CdS composite catalyst with enhanced photocatalytic hydrogen production from water splitting under visible light. Chemical Engineering Journal. 2016;**283**:816-825

[77] Chang CJ, Chu KW, Hsu MH, et al. Ni-doped ZnS decorated graphene composites with enhanced photocatalytic hydrogen-production performance. International Journal of Hydrogen Energy. 2015;**40**:14498-14506

[78] Hou J, Wang Z, Kan W, et al. Efficient visible-light-driven photocatalytic hydrogen production using CdS@TaON core–shell composites coupled with graphene oxide nanosheets. Journal of Materials Chemistry. 2012;**22**:7291

[79] Fan L, Liu PF, Yan X, et al. Atomically isolated nickel species anchored on graphitized carbon for efficient hydrogen evolution electrocatalysis. Nature Communications. 2016; **7**:10667

[80] Kibsgaard J, Jaramillo TF. Molybdenum phosphosulfide: An active, acid-stable, earth-abundant catalyst for the hydrogen evolution reaction. Angewandte Chemie International Edition. 2014;**53**:14433-14437

[81] McKone JR, Sadtler BF, Werlang CA, et al. Ni-Mo nanopowders for efficient electrochemical hydrogen evolution. ACS Catalysis. 2013;**3**:166-169

[82] Tang Y, Hu X, Liu C. Perfect inhibition of CdS photocorrosion by graphene sheltering engineering on TiO_2 nanotube array for highly stable photocatalytic activity. Physical Chemistry Chemical Physics. 2014;**16**:25321-25329

[83] Li H, Xia Z, Chen J, et al. Constructing ternary CdS/reduced graphene oxide/TiO_2 nanotube arrays hybrids for enhanced visible-light-driven photoelectrochemical and photocatalytic activity. Applied Catalysis B: Environmental. 2015;**168-169**:105-113

[84] Lin X, Wang Y, Zheng J, et al. Graphene quantum dot sensitized leaf-like $InVO_4/BiVO_4$ nanostructure: A novel ternary heterostructured QD-RGO/$InVO_4/BiVO_4$ composite with enhanced visible-light photocatalytic activity. Dalton Transactions. 2015;**44**:19185-19193

[85] Zhou J, Tian G, Chen Y, et al. In situ controlled growth of $ZnIn_2S_4$ nanosheets on reduced graphene oxide for enhanced photocatalytic hydrogen production performance. Chemical Communications. 2013;**49**:2237-2239

[86] Sayama K, Yoshida R, Kusama H, et al. Photocatalytic decomposition of water into H_2 and O_2 by a two-step photoexcitation reaction using a WO_3 suspension catalyst and an Fe^{3+}/Fe^{2+} redox system. Chemical Physics Letters. 1997;**277**:387-391

[87] Maeda K, Higashi M, Lu D, et al. Efficient nonsacrificial water splitting through two-step photoexcitation by visible light using a modified oxynitride as a hydrogen evolution photocatalyst. Journal of the American Chemical Society. 2010;**132**:5858-5868

[88] Sasaki Y, Iwase A, Kato H, et al. The effect of co-catalyst for Z-scheme photocatalysis systems with an Fe^{3+}/Fe^{2+} electron mediator on overall water splitting under visible light irradiation. Journal of Catalysis. 2008;**259**:133-137

[89] Iwashina K, Iwase A, Ng YH, et al. Z-schematic water splitting into H_2 and O_2 using metal sulfide as a hydrogen-evolving photocatalyst and reduced graphene oxide as a solid-state electron mediator. Journal of the American Chemical Society. 2015;**137**:604-607

[90] Iwase A, Ng YH, Ishiguro Y, et al. Reduced graphene oxide as a solid-state electron mediator in Z-scheme photocatalytic water splitting under visible light. Journal of the American Chemical Society. 2011;**133**:11054-11057

[91] Suzuki TM, Iwase A, Tanaka H, et al. Z-scheme water splitting under visible light irradiation over powdered metal-complex/semiconductor hybrid photocatalysts mediated by reduced graphene oxide. Journal of Materials Chemistry A. 2015;**3**:13283-13290

[92] Duan L, Bozoglian F, Mandal S, et al. A molecular ruthenium catalyst with water-oxidation activity comparable to that of photosystem II. Nature Chemistry. 2012;**4**:418-423

[93] Sato S, Morikawa T, Saeki S, et al. Visible-light-induced selective CO_2 reduction utilizing a ruthenium complex electrocatalyst linked to a p-type nitrogen-doped Ta_2O_5 semiconductor. Angewandte Chemie International Edition. 2010;**49**:5101-5105

[94] Kumar P, Bansiwal A, Labhsetwar N, et al. Visible light assisted photocatalytic reduction of CO_2 using a graphene oxide supported heteroleptic ruthenium complex. Green Chemistry. 2015;**17**:1605-1609

[95] Kuai L, Zhou Y, Tu W, et al. Rational construction of a CdS/reduced graphene oxide/TiO_2 core–shell nanostructure as an all-solid-state Z-scheme system for CO_2 photoreduction into solar fuels. RSC Advances. 2015;**5**:88409-88413

[96] Gan Z, Wu X, Meng M, et al. Photothermal contribution to enhanced photocatalytic performance of graphene-based nanocomposites. ACS Nano. 2014;**8**:9304-9310

[97] Ali J, Siddiqui G, Yang YJ, et al. Direct synthesis of graphene quantum dots from multilayer graphene flakes through grinding assisted co-solvent ultrasonication for all-printed resistive switching arrays. RSC Advances. 2016;**6**:5068-5078

[98] Zhuo S, Shao M, Lee S-T. Upconversion and downconversion fluorescent graphene quantum dots: Ultrasonic preparation and photocatalysis. ACS Nano. 2012;**6**:1059-1064

[99] Zhao J, Shaygan M, Eckert J, et al. A growth mechanism for free-standing vertical graphene. Nano Letters. 2014;**14**:3064-3071

[100] Davami K, Cortes J, Hong N, et al. Vertical graphene sheets as a lightweight light absorber. Materials Research Bulletin. 2016;**74**:226-233

Chapter 7

CQD-Based Composites as Visible-Light Active Photocatalysts for Purification of Water

Abdullahi Baba Makama, Muneer Umar and
Shettima Abdulkadir Saidu

Additional information is available at the end of the chapter

http://dx.doi.org/10.5772/intechopen.74245

Abstract

The unique physicochemical properties of carbon quantum dot-(CQD)-based photocatalysts, notably their exceptionally good light absorption in the UV and near-visible region, tunable photoluminescence, extraordinary upconversion photoluminescence, outstanding electron affinity, and photoinduced electron transfer, and electron mobility, have attracted considerable attention in different photocatalytic applications. In this review, we summarized the fundamental mechanism and thermodynamics of heterogeneous photocatalysis of aqueous pollutants and the fundamental multifaceted roles of CQDs in photoredox process. Furthermore, we discussed the recent developments in the use of CQD-based materials as visible-light active photocatalysts in water purification. Finally, the challenges and future direction of CQD-based materials as photocatalytic materials for environmental decontamination were highlighted.

Keywords: carbon, CQDs, photocatalyst, upconversion, visible-light

1. Introduction

Carbon quantum dots (CQDs) are a recently discovered class of carbon-based nanomaterials that are typically discrete, quasi-spherical, and less than 10 nm in size [1] (although CQDs of sizes >10 nm have been reported [2–5]). They are composed of sp^2/sp^3-hybridized carbon atoms, have various surface functional groups, and possess composition-dependent fluorescence. This relatively new material has attracted huge interest in many fields including, but not limited to, electrocatalysis [6], biosensing [7–10], bioimaging [11–14], chemical sensing [15], and nanomedicine [16], due to their unique tunable photoluminescence (PL) properties, chemical inertness, high water solubility, ease and low cost of fabrication and, more importantly, low toxicity. Additionally, CQDs have also attracted considerable interest in various photocatalytic applications—environmental remediation [17–21], water splitting to produce H_2 production [22–27], CO_2 conversion [28–30], and synthesis of chemicals [28–33]—because when coupled with a semiconductor photocatalyst, CQDs can provide with several advantages, including improved light harvesting ability, efficient usage of the full spectrum of sunlight, efficient charge carrier separation, stability, and hinder charge recombination. **Figure 1** illustrates the various applications of CQDs in photocatalysis (left) and their competitive optical and structural properties during each photocatalytic procedure (right).

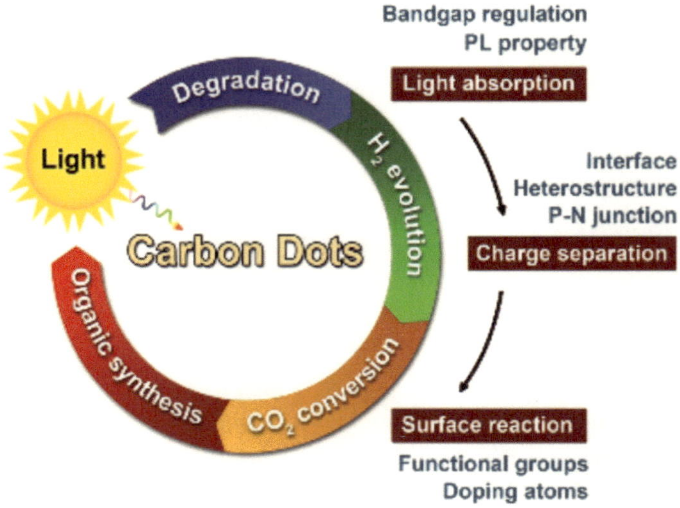

Figure 1. Illustration showing the different areas of applications of CQDs (reproduced from [34]).

2. Principles of heterogeneous photocatalysis

The fundamentals of heterogeneous photocatalysis underlying the degradation of a pollutant employing a semiconductor catalyst have been intensively reported in many literatures [35–41].

A semiconductor (SC) is a material that has a valence band (VB) and a nearly empty conduction band (CB) that are separated from one another by a band gap (E_g). In the ground state (i.e., at $T = 0\ K$), all the electrons are found in the VB of the SC. When illuminated by light of energy greater than or equal to the band gap ($hv \geq E_g$) of a semiconductor, electrons in the VB are excited to the CB, leaving behind equal number of voids (or holes) in the valence band. The generation of electron-hole pairs (e^-_{cb}/h^+_{vb}) is the first step in the heterogeneous photocatalysis of organic and inorganic compounds by semiconductors. This step is illustrated by the enlarged section of **Figure 2**. Once generated, the fate of the electron and hole can follow several de-excitation pathways as shown in **Figure 2**.

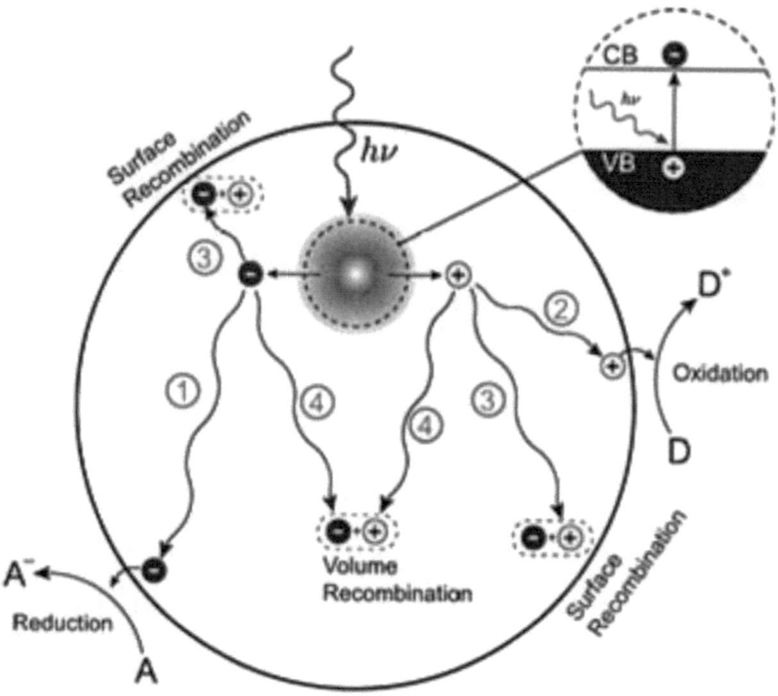

Figure 2. Schematic illustration of the principles and fundamental kinetic requirements of heterogeneous photocatalysis (reproduced from [43]).

First, the spatially separated charge carriers (e^-_{cb}/h^+_{vb}) can migrate to the semiconductor surface where they can induce a redox reaction. At the surface, the electrons can be transferred to reduce an electron acceptor (A) (pathway **1**). On the other hand, a donor species (D) can donate an electron to the surface hole and gets oxidized in the process (pathway **2**). The probability and rate of the charge transfer processes for electrons and holes depends upon the respective positions of the band edges for the conduction and valence bands and the redox potential levels of the adsorbate species [40–42].

In competition with charge transfer to adsorbed species is electron and hole recombination. Recombination of the separated electron and hole can occur on the surface (pathway **3**) or in the volume of the semiconductor particle (pathway **4**). The electron-hole recombination can occur nonradioactively or radioactively to dissipate the input energy either as heat (E) or as a photon ($h\eta$) [44, 45], respectively. The widely postulated steps of heterogeneous photocatalysts may be summarized in a simplified way according to Eqs. (1)–(5).

$$\text{Charge generation}: \ SC + h\eta(\geq E_g) \rightarrow SC(e^-/h^+) \tag{1}$$

$$\text{Charge separation}: \ SC(e^-/h^+) \rightarrow SC(e^-_{cb}, h^+_{vb}) \tag{2}$$

$$\text{Charge recombination}: \ SC(e^-_{vb}, h^+_{cb}) \rightarrow SC + E \text{ or } h\eta \tag{3}$$

$$\text{Charge trapping}: \quad SC(e^-_{vb}, h^+_{cb}) + 2Tr \rightarrow SC(e^-_{Tr}, h^+_{Tr}) \tag{4}$$

$$\text{Reduction reaction} \quad A + e^-_{Tr} \rightarrow A^{\bullet -} \tag{5}$$

$$\text{Oxidation reaction} \quad D + h^+_{Tr} \rightarrow D^{\bullet +} \tag{6}$$

A thermodynamic requirement of spontaneous reaction is a negative change in Gibb's energy, (i.e., $\Delta G < 0$) [46]. In a photocatalytic reaction, this condition is met when the quasi-Fermi (usually taken as the flat band potentials) energy levels of the photogenerated electrons and holes of the photocatalyst straddle the reduction/oxidation potentials of the substrate. In other words, the bottom of the photocatalyst's conduction bands (E^{min}_{CB}) must be located at a more negative potential than the reduction potential (E^0_{red}) of the substrate, while the top of its valence bands (E^{max}_{VB}) must be positioned more positively than the oxidation potential (E^0_{ox}) of the substrate [43, 47] as illustrated in **Figure 3**. Therefore, the spontaneity or otherwise the charge carrier-induced redox reactions, generically depicted by Eq. (5), is determined by the potential of the band edges of the semiconductor photocatalyst [48, 49].

As shown in **Figure 4**, the redox potential of the VB and the CB for different semiconductors varies between +4.0 and −1.5 V vs. normal hydrogen electrode (NHE), respectively. Many organic compounds have more negative oxidation potentials than the valence potential of most semiconductor photocatalysts. Accordingly, it is thermodynamically possible for the organic pollutants to be oxidized by photocatalysts. In contrast, fewer organic compounds can be reduced by photocatalyst since only a smaller number of them have a potential below that of the conduction band of most photocatalysts [50].

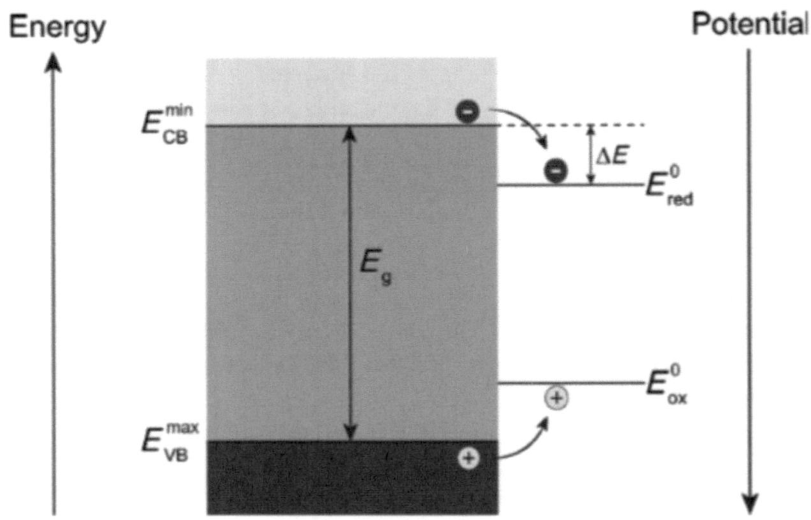

Figure 3. Thermodynamic constraints on the transfer of charge carriers to adsorbed molecules, and ΔE represents the kinetic overpotential of the reduction process. In this scheme, electrons and holes gain stability when moving down and up, respectively (reproduced from [43]).

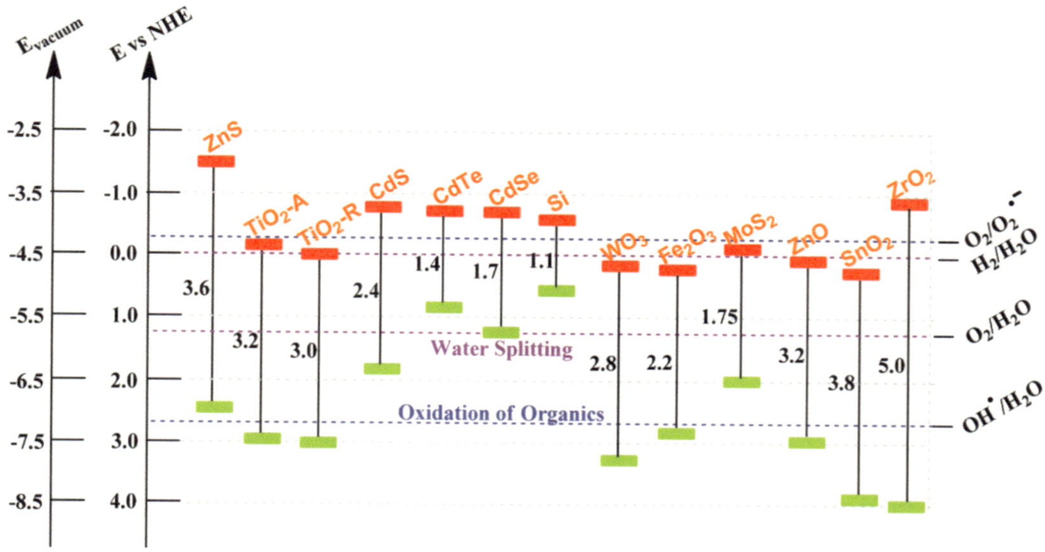

Figure 4. Bandgap energy structure of several common semiconductors on a potential scale (V) vs. NHE/vacuum (reproduced from [51]).

While choosing a photocatalyst for application in environmental remediation, it is also essential to take into account the reduction potentials (E) of the substrate, as well as those of the intermediates that are formed during the photocatalytic reaction(s). However, the reduction potentials for different organic or inorganic compounds are usually dependent on the reaction conditions such as pH and the employed electrolyte. The reduction potential of the couples $M/M^{\bullet -}$ refers to the reactions described by Eqs (7) and (8):

$$\text{oxidant} + e^{-} \rightarrow \text{Reductant} \qquad (7)$$

$$\text{oxidant} + e^{-} + nH^{+} \rightarrow \text{Reductant} \qquad (8)$$

These reactions refer to one-electron reduction reactions—the potentials of which are given versus the standard hydrogen electrode (SHE). **Table 1** includes the reduction potentials of the most important inorganic species that may be present in photocatalytic systems.

For example, over a TiO_2 photocatalyst in pH 7 solution, reaction Eq. (9) is feasible because the redox potential of the $e_{cb}^{-} = -0.52$ V vs. SHE is sufficiently more negative than the reduction potential ($E^0(O_2/O_2^{\bullet -})$ = -0.28 V vs. SHE) of the superoxide radicals. Reaction (10) occurs because the redox potential of the $h_{vb}^{+} = +2.53$ V vs. SHE is sufficiently more positive than the oxidation potential ($E^0(H_2O/OH^{\bullet}) = +2.27$ V vs. SHE) of the hydroxyl radical.

$$O_2 \text{ isonosorption}: \qquad O_2 + e_{cb}^{-} \rightarrow O_2^{\bullet -} \qquad (9)$$

$$H_2O \text{ ionization}: \qquad H_{ads}^{+} + OH_{ads}^{-} + h_{vb}^{+} \rightarrow OH^{\bullet} + H^{+} \qquad (10)$$

Reductiom reaction	E^a, (V)	Remark
$aq + e^- \rightarrow e^-_{aq}$	−2.870	
$SO_4^{2-} + H_2O + e^- \rightarrow SO_3^{\bullet -} + 2OH^-$	−2.470	
$CO_2 + e^- \rightarrow CO_2^{\bullet -}$	−1.900	
$O_2 + e^- \rightarrow O_2^{\bullet -}$	−0.330	
$O_2 + H^+ + e^- \rightarrow HO_2^{\bullet -}$	−0.037	pH 0
$Cl_2 + e^- \rightarrow Cl_2^{\bullet -}$	0.420–0.600	
$I_2 + e^- \rightarrow I_2^{\bullet -}$	0.420–0.600	
$SO_3^{\bullet -} + e^- \rightarrow SO_3^{2-}$	0.630	pH >8
$Fe^{3+} + e^- \rightarrow Fe^{2+}$	0.570	
$HO_2^{\bullet} + e^- \rightarrow HO_2^-$	0.590	
$H_2O_2 + H^+ + e^- \rightarrow H_2O + HO^{\bullet}$	0.800	pH 7
$Ag^+ + e^- \rightarrow Ag$	0.800	
$NO_2^{\bullet} + e^- \rightarrow NO_2^-$	0.870–1.040	
$O_2^{\bullet -} + 2H^+ + e^- \rightarrow H_2O_2$	0.940	pH 7
$O_2^{\bullet -} + H^+ + e^- \rightarrow HO_2^-$	1.000	
$CO_3^{\bullet -} + H^+ + e^- \rightarrow HCO_3^-$	1.070	pH 7
$HS^{\bullet} + e^- \rightarrow HS^-$	1.150	
$O_3 + e^- \rightarrow O_3^{\bullet -}$	1.190–1.600	pH >11
$I^{\bullet} + e^- \rightarrow I^-$	1.270–1.420	
$HO_1^{\bullet} + H^+ + e^- \rightarrow H_2O_2$	1.420	pH 9
$CO_3^{\bullet -} + e^- \rightarrow CO_3^{2-}$	1.500	
$O_3 + H^+ + e^- \rightarrow HO_3^{\bullet}$	1.800	pH 7
$OH^{\bullet} + H^+ + e^- \rightarrow H_2O$	1.800–2.180	pH 7
$CN^- + e^- \rightarrow CN^-$	1.900	
$Br^- + e^- \rightarrow Br^-$	2.000	
$Cl^{\bullet} + e^- \rightarrow Cl^-$	2.200–2.600	
$Cl_2^{\bullet -} + e^- \rightarrow 2Cl^-$	2.300	
$NO_3^{\bullet} + e^- \rightarrow NO_3^-$	2.300–2.600	
$SO_4^{\bullet -} + e^- \rightarrow SO_4^{2-}$	2.430	
$OH^{\bullet} + H^+ + e^- \rightarrow H_2O$	2.590–2.850	pH 0
$F^{\bullet} + e^- \rightarrow F^-$	3.600	

[a]Reduction potential referring to one-electron reduction vs. SHE in the case of proton involvement, and E values may be used to estimate the standard potentials E^o as defined by the Nernst equation.

Table 1. Reduction potentials of some species that may be involved in photocatalytic systems (adapted from [51]).

Moreover, it is in fact critical to provide an overpotential for each process, to initiate and subsequently drive the electron transfer process. Without an overpotential, even a good photocatalyst cannot ensure sufficiently a high rate of reaction.

3. Beneficial roles of CQD materials in photocatalysis

3.1. Extends the optical absorption range of photocatalysts

The absorption of large portion of incident solar radiation by a photocatalyst is one of the key factors for achieving useful efficiencies in photocatalysis [52]. For wide bandgap semiconductors such as TiO_2, a common strategy for achieving this is through the use of photosensitizers. A photosensitizer is excited by low-energy photons to generate electrons that are subsequently injected into the conduction band of the wide bandgap photocatalyst. The common photosensitizers are usually characterized as expensive [53–55], toxic [56], unstable [57], and polluting [58]. These drawbacks limit the practical and large-scale application of conventional photosensitizers. Accordingly, it is essential to find an alternative photosensitizer material(s) that is free from these drawbacks and one that can harvest larger portion of the solar radiation. Carbon quantum dot appears to be such a material. It is easy to produce, cheap, nontoxic, green, stable, and abundant [1, 58, 59]. Additionally, it displays strong blue photoluminescence and good optical absorption in the UV and near-visible region [60]. Futhermore, CQDs exhibit upconversion photoluminescence properties [19, 61–63] and exhibit promising electron transfer properties [64, 65]. These optical characteristics of CQDs render them promising candidates as photosensitizers for photocatalytic applications. Three possible photsensitization mechanisms are proposed for CQDs in the literature. Firstly, CQD can be excited by a low-energy radiation to generate holes (h^+_{vb}) and electrons (e^-_{cb}). Under favorable thermodynamic condition, the photogenerated e^-_{cb} is injected into the conduction band of a wide bandgap photocatalyst (such as TiO_2) to initiate a reaction as shown in **Figure 5(a)** [19]. Secondly, the CQDs with upconversion PL properties can convert longer wavelength light to the short wavelength light, which in turn can excite a wide bandgap photocatalyst (ZnO) to form e^-_{cb}/h^+_{vb} pairs (**Figure 5(b)** [66]). Finally, addition of CQDs may lead to bandgap narrowing of the semiconductor owing to the chemical bonding between semiconductor and CQD, which results in the extended light absorption range as shown in **Figure 5(c)** [67].

Recently, several reports have been published that show CQDs to dramatically extend the optical absorption range of wide and narrow bandgap photocatalysts to the entire visible-light range and beyond [18–21, 58, 60, 62, 68–71]. For example, in a recent publication, Ye et al. [58] reported using CQDs to extend the light absorption range of $BiVO_4$ to the entire visible range. The resulting CQD/$BiVO_4$ composite photocatalyst achieved an absolute photocurrent density J_{abs} of 9.2 mA cm^{-2} as shown in **Figure 6(a)**. A photoanode fashioned from the composite (NiOOH/FeOOH/CQDs/$BiVO_4$) achieved a photocurrent density of 5.99 mA cm^{-2} at 1.23 V vs. RHE under AM 1.5G in KH_2PO_4 aqueous solution without a hole scavenger (pH 7) and record a high applied bias photon-to-current efficiency of 2.29% at 0.6 V vs. RHE. Ren et al. [18]

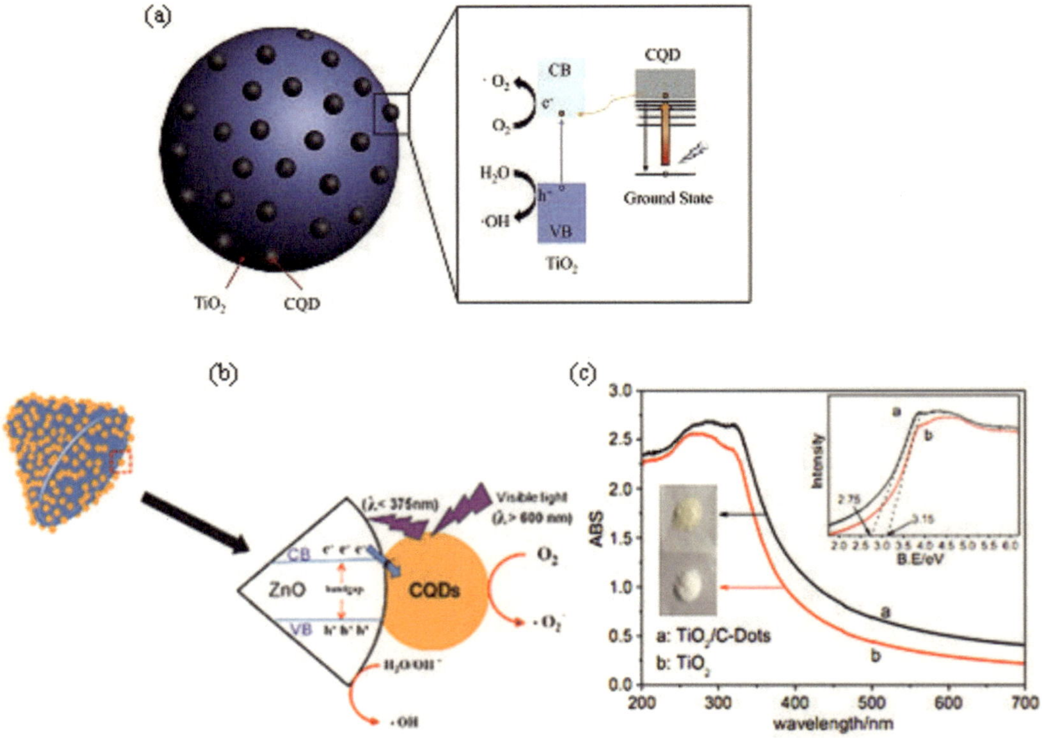

Figure 5. (a) Schematic illustration of the sensitization mechanism of CQDs based on (a) photoexcitation of CQDs [19], (b) upconversion photoluminescence [66], and (c) narrowing of band gap [67].

reported that coupling of CQDs with $MoSe_2$ resulted in a composite, CQD/$MoSe_2$ photocatalyst that is active over the entire range of the solar spectrum (**Figure 6(b)**).

Carbon quantum dots have also been applied as visible-light photosensitizer for non-noble metal H_2-evolution catalyst [24, 60, 68]. Martindale et al. used CQDs as a photosensitizer to energize a nickel-based molecular catalyst, Ni-bis-(diphosphine) (NiP) to produce H_2 under visible-light irradiation [24, 60]. A proposed scheme of H_2 production in the homogeneous CQD-NiP system is presented in **Figure 6(c)**. Irradiation of photoluminescent CQDs with visible-light results in the direct transfer of photoexcited electrons to the catalyst NiP with subsequent reduction of aqueous protons to H_2. The electron donor EDTA [60] or TCEP/EDTA [24] quenches the photogenerated holes in the CQDs. In a similar work, McCormick and co-workers [68] used a PVP-coated CQD as a photosensitizer for a nickel nanoparticle (NiNP) catalyst (**Figure 6(d)**) to produce H_2 at a much higher quantum yield of 6%. In this report, it was observed that the fluorescence quantum yield of CQDs increases with increasing PVP coating. However, H_2 production decreased when the PVP coating of CQD is greater than 20%. A total of 330 H_2/g CQD was collected from a 20% PVP-coated CDQ/Ni nanoparticle system at a pH of 4.51 after five irradiation with a 470-LED light source.

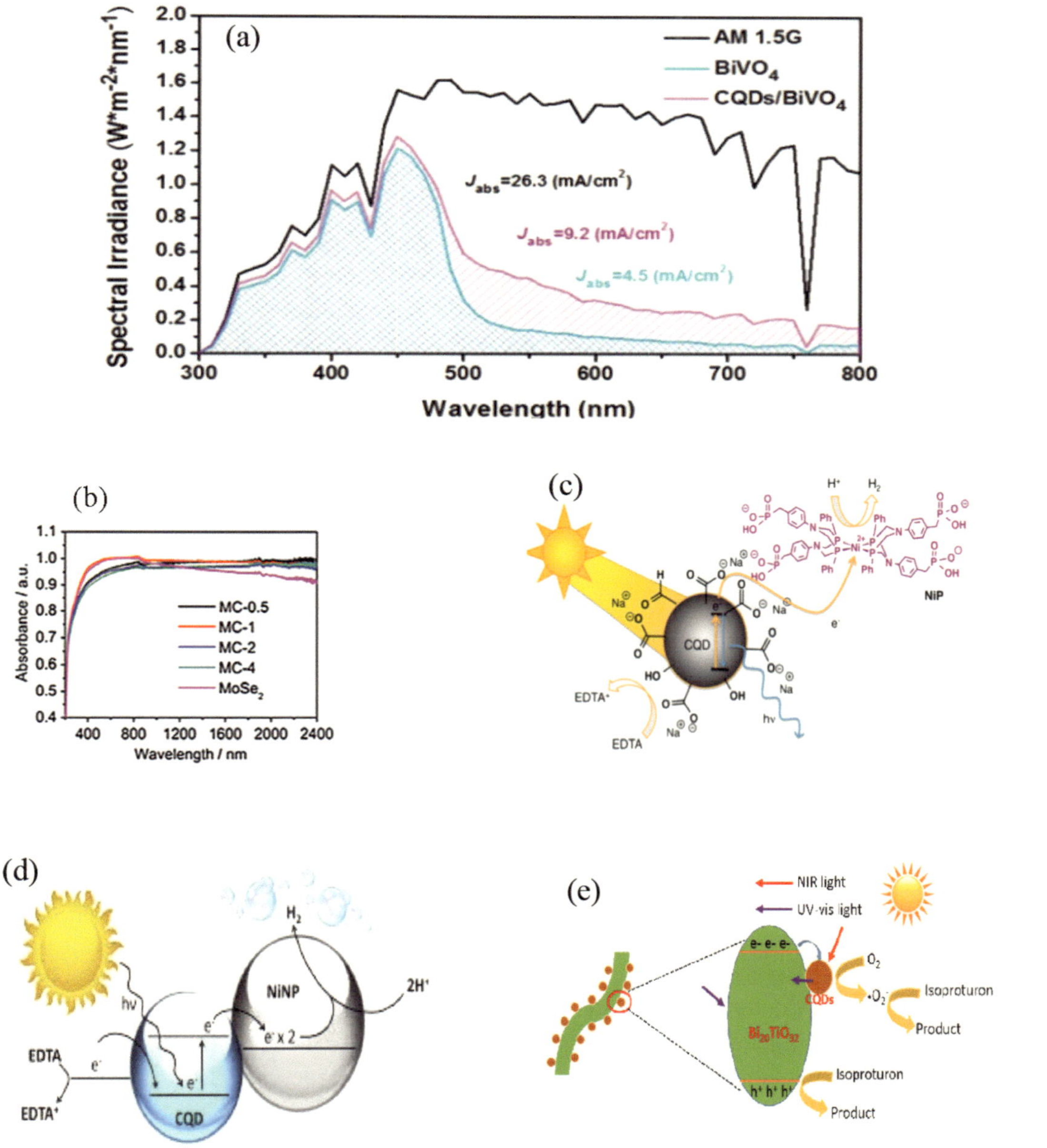

Figure 6. (a) Spectrum of the solar irradiance of AM 1.5G (ASTM G173-03) and the spectra showing the efficiencies of light absorption (LHE) of the $BiVO_4$ and CQDs/$BiVO_4$ photoanodes at 300–800 nm [58]. (b) Diffuse absorption spectra of $MoSe_2$ and the CQD/$MoSe_2$ (denoted as MC- in the plot) composites with different CQD contents [18]. Photosensitization of metal catalyst: (c) molecular NiP catalyst [60] and (d) nickel nanoparticle (NiNP) catalyst [68] under visible-light irradiation. (e) Sensitization of $Bi_{20}TiO_{32}$ by upconversion photoluminescence property of CQDs under NIR radiation [62].

Upconversion photoluminescence property (UCPL) of CQDs was used to improve the visible-light and NIR response of narrow bandgap composite photocatalysts. For example, Liu and coworkers [72] used the upconversion property of nitrogen-doped CQDs to improve the visible-light and near infra-red response of graphitic carbon nitride (g-C_3N_4). Similarly, Mao et al. [62] improved the visible-light and NIR response of $Bi_{20}TiO_{32}$ by coupling it with CQDs. The resulting composite photocatalyst exhibited improved photoactivity under NIR irradiation. **Figure 6(e)** shows the proposed CQD sensitization mechanism.

3.2. Enhances charge separation and electron transfer

One of the main processes that limit the quantum efficiency of a photocatalytic system is the fast recombination of photogenerated charge carriers [40]. Therefore, to enhance the performance of a photocatalyst, it is essential to improve charge carrier separation and minimize the rate of their recombination. Several approaches have been devised to achieve this goal. These included strategies such as surface modification of the semiconductor particles with noble metals [73, 74], coupling of two semiconductor particles with different electronic levels [75–77], and using sacrificial reagents to scavenge for photogenerated electrons or holes [78–80]. Another strategy is to couple carbon-based materials with a photocatalyst because of their high charge storage capacity and electrical conductivity [81–83]. In particular, many reports have been recently published that show the beneficial effect of coupling of CQDs with photocatalysts [18–21, 63, 71, 84–86]. The intrinsic band gap and strong electron affinity [59, 65, 87] of CQDs give them the ability to readily accept photogenerated electrons from an electron donor such as a semiconductor with a more negative conduction band minimum. The transferred electrons are then shuttled freely along the conducting paths of the CQDs allowing for effective charge separation, stabilization, and prevention of charge recombination. The longer-lived charge carriers have greater probability to induce transformations, thus accounting for the much improved quantum efficiency of the CQD-based photocatalysts. **Figure 7** depicts a proposed photoinduced electron transfer mechanism on a layered composite of CQD/CdS photocatalyst [88].

Results of time-resolved (TR) PL and photocurrent measurements corroborated the slow recombination rates and enhanced the separation of electron-hole pairs observed in CQD-based photocatalysts. **Figure 7(b)** and **(d)** showed the time-resolved photoluminescence (TRPL) measurements on pure $Bi_{20}TiO_{32}$ and CdSe and their CQD-based counterparts with different amounts of CQD. Because the TRPL plots of the CQD composite photocatalysts exhibited lower PL intensities, it implies that the CQDs' modification effectively inhibits the recombination of photogenerated electron-hole-electron pairs in the hybrid photocatalysts. **Figure 7(c)** and **(e)** showed the corresponding photocurrent responses. The higher photocurrent signals recorded for the CQD-based photocatalysts compared to the pure photocatalysts demonstrated that much more photogenerated charge carriers were produced and the electron-hole pairs could separate more efficiently [62, 89].

3.3. Provides additional surface for adsorption and reaction

Another key role of CQD, which promotes photocatalytic activity, is its capacity to provide additional surface for adsorption and reaction of substrates. As a nanomaterial, CQDs possess a larger surface area to volume ratio, and thus, its composite with other nanomaterials potentially

http://dx.doi.org/10.5772/intechopen.74245

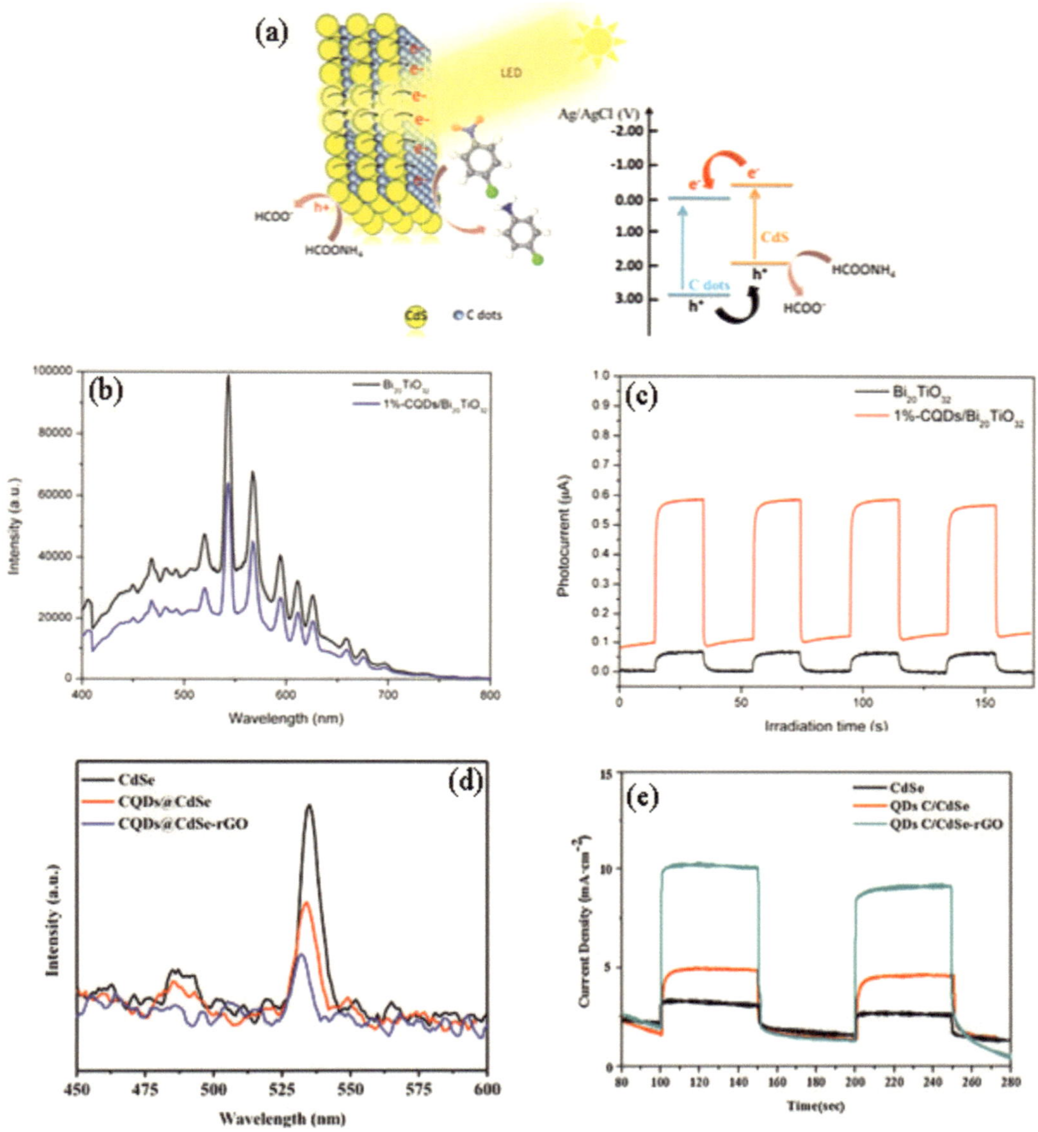

Figure 7. (a) Illustration of the mechanism of photoinduced electron transfer on a layered CQD/CdS nanocomposite photocatalyst from Ref. [88]. Plots (b)–(e) show the time-resolved photoluminescence and transient photocurrent profiles of pure photocatalysts and their CQD-based counterparts. Charts (b) and (c) for $Bi_{20}TiO_{32}$ and CQD/$Bi_{20}TiO_{32}$ from [62], and (d) and (e) pure for CdSe and CQD-based photocatalysts from Ref. [89].

acquires much enhanced surface area, thereby increasing the adsorption capacity. Like other carbon materials, adsorption capacity of CQD originates from the flexible sp^2-bonded carbon structure and the large surface area. Because the surface of CQDs contains a plethora of oxygenated functional groups [59], its adsorption capability depends on its interaction with the adsorbate. The aromatic regions of CQDs can form a π-π stacking interaction with organic adsorbates containing aromatic structures to enhance their adsorption. On the other hand, the presence of functional groups such as carbonyl, epoxy, hydroxyl, and amino groups on the surface of CQDs promotes the adsorption of a wide variety of molecules and metal ions. The increased adsorption of target reactants on the surface of photocatalyst increases their chance of reacting with photogenerated reactive oxidative species, thus enhancing photocatalytic activity of the composite CQDs-based photocatalyst [90–92].

4. CQD-based semiconductor composites as visible-light active photocatalysts for water treatment

Recently, a large number of studies [17–21, 85, 86, 93–95] have reported on the application of CQD/semiconductor nanocomposites as photocatalysts for water purification owing to the exceptional properties exhibited by the composites. Carbon quantum dots have been coupled with variety of metal-oxide semiconductors such as TiO_2, Ag_3PO_4, SiO_2, $MoSe_2$, Bi_2MoO_6, and others and have found to be an excellent photocatalyst for the degradation of pollutants. **Table 2** summarizes some of the CQD-based composites that have been investigated as photocatalysts for the degradation of aqueous pollutants.

Chen et al. [17] reported the fabrication of nitrogen doped (i.e., NCQD/Ag_3PO_4) complex photocatalyst with improved activity under visible-light irradiation. As shown in **Figure 8(a)**, the activities of the hybrid NCQD/Ag_3PO_4 materials are significantly enhanced for the degradation of methyl orange (MO) compared with the pure Ag_3PO_4 under visible light ($\lambda >$ 420 nm). A composite of NCQD/Ag_3PO_4 prepared from a 3 mL solution of 0.1 g CQD/L of distilled water denoted 3-NCQD/Ag_3PO_4 exhibited the highest activity. The 3-NCQD/Ag_3PO_4 sample eliminated 98% of the MO present in 18 min. In another report, Ren et al. [18] reported

CQD composite	Targeted pollutant	Visible-light source	Experiment conditions	η, %	Reference
CQD/Bi_2MoO_6	Methylene blue (MB)	300 W Xe lamp, ≤ 400-nm filter	100 mg cat., MB (100 mL, 10 mg/L), 30 C, 120 min	—	[20]
CQDs/Ag/Ag_2O	Methylene blue	250 W Xe lamp, ≤ 420-nm filter	50 mg cat., MB (100 mL, 10 mg/L), 150 min	≈ 100	[94]
CQD-TiO_2	Methylene blue	500 W Xe lamp, ≤ 420-nm filter	MB (15 mL, 5 mg/L), 100 min	35	[85]
CQD-TiO_2	Methylene blue	> 420 nm light	MB (50 mg/L), 25 min	≈ 100	[86]
CQD-TiO_2	Methylene blue	> 420 nm light	120 min	90	[19]
CQD-SiO_2	Methylene blue	> 420 nm light	MB (50 mg/L), 15 min	≈ 100	[86]
Fe_2O_3-CQD	Methylene blue	400 W halogen, ≤ 400-nm filter	50 mg cat., MB (100 mL, 20 mg/L), +35% H_2O_2, 90 min	97.3	[96]
Cu_2O-CQD	Methylene blue	400 W halogen, ≤ 400-nm filter	30 mg cat., MB (100 mL, 50 mg/L), 120 min	88	[97]

CQD composite	Targeted pollutant	Visible-light source	Experiment conditions	η, %	Reference
N-CQD	Methylene blue	> 420 nm light	10 μL cat., MB (3 mL, 1×10^{-4}mol dm^3),+$NaBH_4$, 6.5 min	≈ 100	[93]
Bi_2MoO_6-CDQ	Ciprofloxacin (CIP)	300 W Xe lamp with 400-nm filter	100 mg cat., CIP (100 mL, 10 mg/L), 30°C, 120 min	88	[20]
CDQ/Bi_2WO_6	Ciprofloxacin	300 W Xe lamp, ≤ 400-nm filter	50 mg cat., CIP (100 mL, 10 mg/L), 120 min	87	[21]
CQD/BiOBr	Ciprofloxacin	300 W Xe lamp, ≤ 400 nm filter	30 mg cat., CIP (100 mL, 10 mg/L), 4 h	44.3	[95]
Bi_2MoO_6-CDQ	Bisphenol A (BPA)	300 W Xe lamp, 400-nm filter	100 mg cat., BPA (100 mL, 10 mg/L), 30°C, 120 min	54	[20]
CQDs/BiOI	Bisphenol A	300 W Xe lamp, ≤ 420-nm filter	30 mg, 50 mg, 70 mg Cat., BPA (100 mL, 10 mg/L), 120 min	99	[84]
Bi_2WO_6-CDQ	Bisphenol A	300 W Xe lamp, ≤ 400-nm filter	50 mg cat., BPA (100 mL, 10 mg/L), 120 min	–	[21]
CQD/BiOBr	Bisphenol A	300 W Xe lamp, ≤ 400-nm filter	50 mg cat., BPA (100 mL, 10 mg/L),	66	[95]
Bi_2MoO_6 – CDQ	Tetracycline chloride (TC)	300 W Xe lamp, 400-nm filter	100 mg cat., TC (100 mL, 20 mg/L), 30, 120 min	–	[20]
CQDs/BiOI	Tetracycline chloride	300 W Xe lamp, ≤ 420-nm filter	30 mg, 50 mg, 70 mg Cat., TC (100 mL, 20 mg/L), 120 min	68	[84]
Bi_2WO_6 – CDQ	Tetracycline chloride	300 W Xe lamp, ≤ 400-nm filter	50 mg cat., TC (100 mL, 20 mg/L), 120 min	–	[21]
$MoSe_2$ – CDQ	Cr^6+	> 420 nm light	180 min	99	[18]
Ag_3PO_4 – CQD	Methyl orange (MO)	150 W Xe lamp, ≤ 420-nm filter	200 mg cat., MO (100 mL, 66 ppm), 25 min	≈ 100	[70]
CQD – TiO_2	Rhodamine B (RhB)	500 W Xe lamp, ≤ 420-nm filter	100 mg cat., MB (100 mL, 10 ppm), 60 min	95.4	[71]
CQDs/BiOI	Rhodamine B	300 W Xe lamp, ≤ 420-nm filter	30 mg, 50 mg, 70 mg Cat., RhB (100 mL, 10 mg/L), 30 min	61	[84]
CQD/BiOBr	Rhodamine B	300 W Xe lamp, ≤ 400-nm filter	20 mg cat., RhB (100 mL, 10 mg/L), 30 min	≈ 100	[95]
CQDs/Ag/Ag_2O	Rhodamine B	150 W NIR lamp, ≤ 700-nm filter	50 mg cat., RhB (100 mL, 10 mg/L), 150 min	48	[94]
Bi_2WO_6-CDQ	Rhodamine B	300 W Xe lamp, ≤ 400-nm filter	50 mg cat., RhB (100 mL, 10 mg/L), 120 min	–	[21]
CQDs/BiOI	Rhodamine B	300 W Xe lamp, ≤ 400-nm filter	30 mg cat., RhB (100 mL, 10 mg/L), 120 min	94.9	[98]
CQDs/CdS	Nitro-benzene	3 W LED light, ≤ 420-nm filter	30 mg cat., RhB (15 mL, 10 mg/L), 120 min, +$HCOONH_4$	–	[88]

Table 2. Different CQD-based composites used as photocatalysts used for water purification under visible-light irradiation.

improved activity for the reduction of Cr(VI) ion over CQD/$MoSe_2$ composite photocatalyst compared to the activity over pure $MoSe_2$ (**Figure 8(b)**). Similarly, Xie et al. also reported on the beneficial effects of complexing CQDs with $Bi_{20}TiO_{32}$ photocatalysts. Visible light active to improve the rate of charge transfer and reducing the rate of h^+_{vb}-e^-_{cb} recombination in $Bi_{20}TiO_{32}$ photocatalysts. The resulting CQD/$Bi_{20}TiO_{32}$ composites exhibited improved photocatalytic activities as shown in **Figure 8(c)**.

Li et al. [86] first reported the exploitation of the unique optical properties of CQDs in photocatalysis. The team synthesized TiO_2/CQDs and SiO_2/CQDs nanocomposite systems (**Figure 9a**) from CQDs with upconversion photoluminescence (PL) properties (**Figure 9b**) and used them to investigate the degradation of aqueous methylene blue (MB) under visible light. The as-synthesized CQDs-based photocatalysts exhibited enhanced ability to promote the degradation of the dye compared to their precursor components (CDQs, TiO_2, or SiO_2) as shown in **Figure 9c**. Under the given experimental conditions (50 mgL^{-1} MB solution, suspended TiO_2/CQDs or SiO_2/CQDs, 300 W halogen lamp irradiation) used, the MB was almost completely degraded in 15 and 25 min over SiO_2/CQDs and TiO_2/CQDs, respectively.

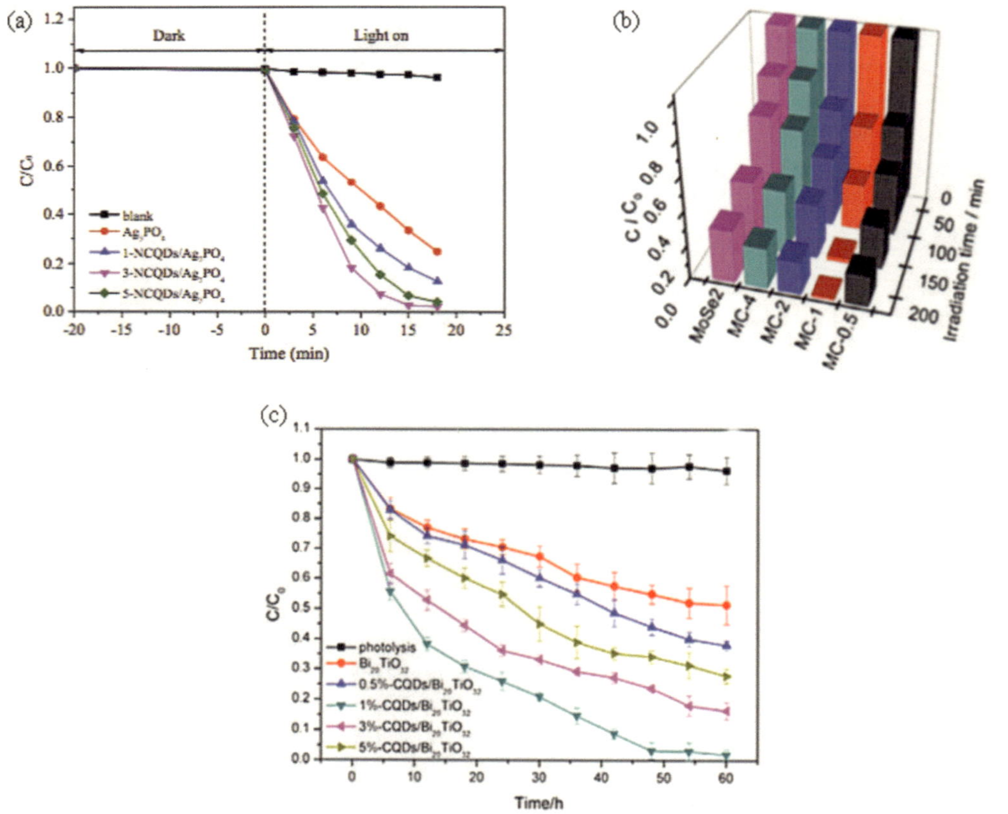

Figure 8. Photocatalytic performance of (a) MO over NCQD/Ag_3PO_4. (b) Cr(VI) over CQD/$MoSe_2$. (c) Isoproturon over CQD/$Bi_{20}TiO_{32}$.

Li and co-workers attributed the enhanced photocatalytic activities of the nanocomposites under visible light to upconversion and efficient charge carrier separation in the nanocomposite systems. The former allows CQDs to emit short-wavelength photons after absorbing visible light. The emitted short-wavelength UV light then induces photoexcitation of the TiO_2 and SiO_2 to generate the electron/hole (e^-_{cb}/h^+_{vb}) redox pairs that go to initiate the process of MB degradation via the generation reactive oxygen species such as $O_2^{\bullet -}$ and $OH^{\bullet}$ and others. The appropriate position of the conduction band edge of the CQDs allows electron transfer from the surface of the semiconductors, thereby enhancing charge carrier separation and inhibiting recombination resulting in the increased number of e^-_{cb} and h^+_{vb} available for photoreactions.

Yu et al. [96] reported the synthesis of CQDs/Fe_2O_3 complex photocatalysts via a facile solvent-thermal process in an aqueous solution (**Figure 10a**). The photocatalytic activity of photocatalyst was evaluated by the degradation of MB as a model contaminant under visible-light irradiation. The results (**Figure 10b**) showed that coupling of CQDs with Fe_2O_3 greatly enhances the

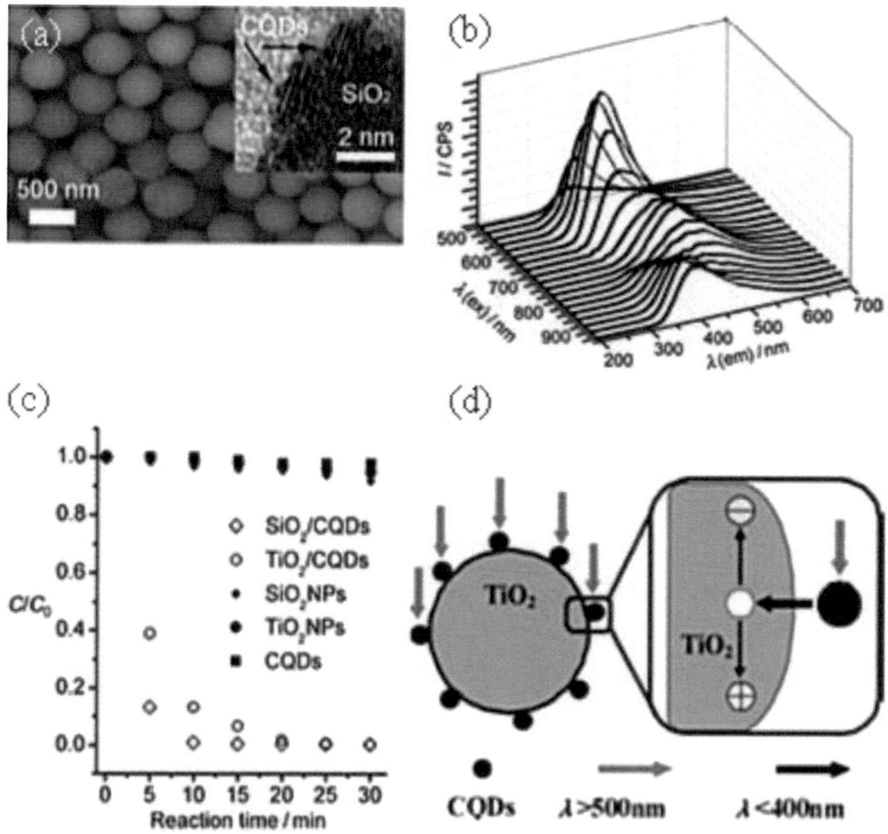

Figure 9. (a) A representative SEM image of the SiO_2/CQDs nanocomposite photocatalyst. Inset shows the corresponding HRTEM image. (b) Excitation-emission profile showing the upconversion PL property of the as-synthesized CQDs. (c) Degradation profile of MB in suspensions of different CQDs-based photocatalysts and control samples. (d) Scheme showing the postulated explanation of the mechanism of photocatalysis over CQDs-based photocatalyst. TiO_2/CQDs are used as a representative (reproduced from [86]).

decomposition of MB under visible-light irradiation. The authors attributed the enhanced activity of the photocatalyst to the mesoporous architecture and large surface area of the α-Fe_2O_3, which provides more abundant active sites for the adsorption and degradation of MB to occur and also for the effective electron transfer from the conduction band of the Fe_2O_3 particle to the conducting networks of the loaded CQDs as shown in **Figure 10c**.

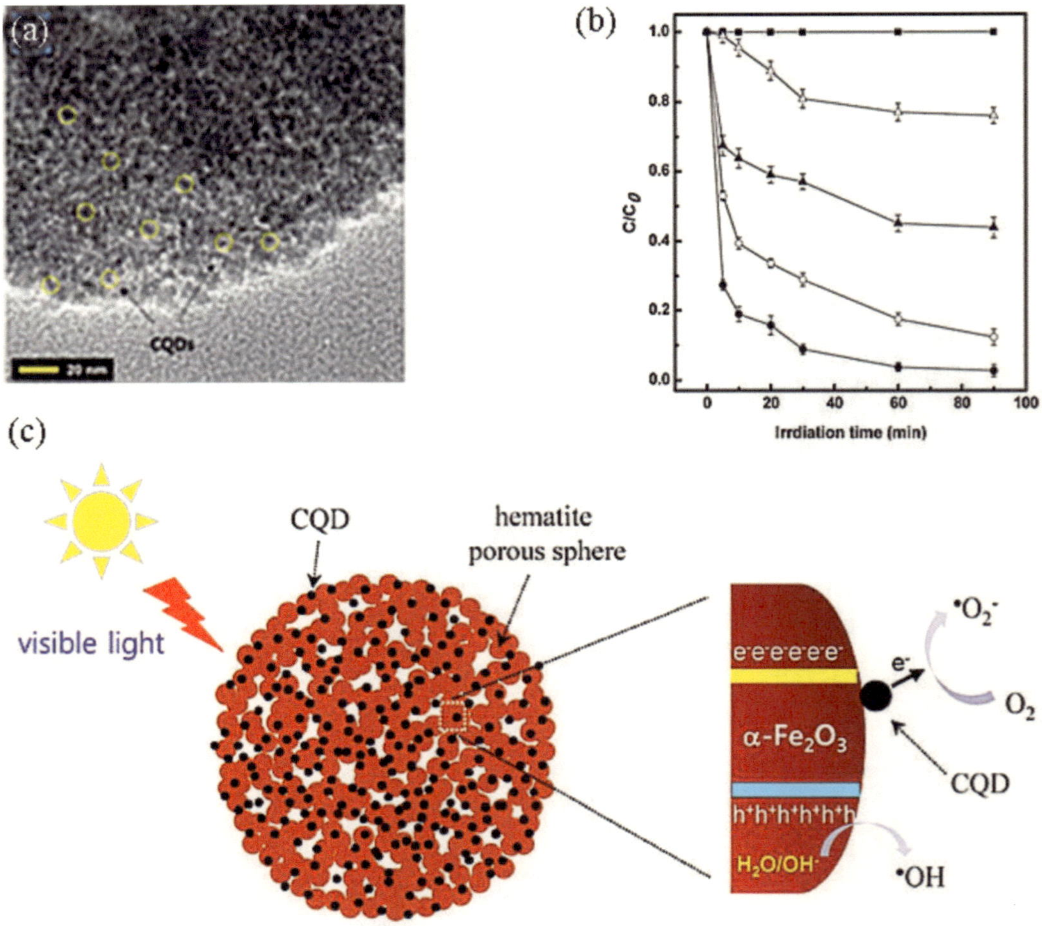

Figure 10. (a) HRTEM image of the CQDs/Fe_2O_3 photocatalysts. (b) Comparison of MB degradation profile in the presence of different catalysts: (■) CQD/Fe_2O_3 without light, (Δ)Fe_2O_3, (▲)$Fe_2O_3 + H_2O_2$, (O) CQD/Fe_2O_3, and (•) CQD/$Fe_2O_3 + H_2O_2$ (reproduced from [96]).

5. Conclusions and outlook

In summary, various types of carbon quantum dots/semiconductor composites have been investigated for the design of novel materials for water remediation via heterogenous photocatalysis. The coupling of CQDs with semiconductors synergistically enhances the activity of the photocatalysts for the degradation of various waterborne pollutants. The unique properties of CQDs, in conjunction with size-dependent properties of nanomaterials, induce further functionalities to the composites such as high adsorption capacity, extended light absorption range, and improved charge separation properties along with high stability. Despite the confirmation of the excellent activity of CQDs-based photocatalystsic aforementioned reports, the applicability of the heterogeneous photocatalytic technology in large-scale water treatment is constrained by several key technical issues that have to be addressed.

The main constraints associated with the field scale application of CQD-based photocatalysts essentially lies in the design of (i) a catalyst that can utilize large portion of the solar spectra, (ii) catalyst immobilization strategy to provide a cost-effective solid-liquid separation, (iii) improvement in the photocatalytic operation for wider pH range and to minimize the addition of oxidant additives, (iv) new integrated or coupling system for enhanced photomineralization or photodisinfection kinetics, and (v) effective design of photocatalytic reactor system or parabolic solar collector for higher utilization of solar energy to reduce the electricity costs.

Acknowledgements

The authors gratefully acknowledge the financial support of the Tertiary Education Trust Fund of the Nigeria Ministry of Education.

Author details

Abdullahi Baba Makama[1]*, Muneer Umar[1] and Shettima Abdulkadir Saidu[2]

*Address all correspondence to: abmakama@hotmail.com

1 Department of Chemical Engineering Technology, The Federal Polytechnic, Nasarawa, Nigeria

2 Department of Science Laboratory Technology, The Federal Polytechnic, Nasarawa, Nigeria

References

[1] Baker SN, Baker GA. Luminescent carbon nanodots: emergent nanolights. Angewandte Chemie (International Ed. in English). 2010;**49**(38):6726-6744. DOI: 10.1002/anie.200906623. ISSN: 1521-3773 (Electronic); 1433-7851 (Linking)

[2] Yang P, Zhao J, Wang J, Cui H, Li L, Zhu Z. Multifunctional nitrogen-doped carbon nanodots for photoluminescence, sensor, and visible- light-induced H_2 production. ChemPhysChem. 2015;**16**(14):3058-3063. DOI: 10.1002/cphc.201500447. ISSN: 1439-7641 (Electronic); 1439-4235 (Linking)

[3] Yang P, Zhao J, Wang J, Cao B, Li L, Zhu Z. Construction of Z-scheme carbon nanodots/ WO_3 with highly enhanced photocatalytic hydrogen production. Journal of Materials Chemistry A. 2015;**3**(16):8256-8259. DOI: 10.1039/C5TA00657K

[4] Liu J, Zhu W, Yu S, Yan X. Three dimensional carbogenic dots/TiO_2 nanoheterojunctions with enhanced visible light-driven photocatalytic activity. Carbon. 2014;**79**:369-379. DOI: 10.1016/j.carbon.2014.07.079. URL: http://www.sciencedirect.com/science/article/pii/S0008622314007301

[5] Wang H, Wei Z, Matsui H, Zhou S. Fe_3O_4/carbon quantum dots hybrid nanoflowers for highly active and recyclable visible-light driven photocatalyst. Journal of Materials Chemistry A. 2014;**2**(38):15740-15745. DOI: 10.1039/C4TA03130J

[6] Shen J, Li Y, Su Y, Zhu Y, Jiang H, Yang X, Li C. Photoluminescent carbon-nitrogen quantum dots as efficient electrocatalysts for oxygen reduction. Nanoscale. 2015;**7**(5): 2003-2008. DOI: 10.1039/c4nr06484d. ISSN: 2040-3372 (Electronic); 2040-3364 (Linking)

[7] Garg B, Bisht T. Carbon nanodots as peroxidase nanozymes for biosensing. Molecules. 2016;**21**(12). DOI: 10.3390/molecules21121653. ISSN: 1420-3049 (Electronic); 1420-3049 (Linking)

[8] Hao J, Wang L, Wang Y, Sun X, Zhang G, Dong S. Versatile self-assembly and biosensing applications of DNA and carbon quantum dots coordinated cerium ions. Chemistry. 2017. DOI: 10.1002/chem.201701709. ISSN: 1521-3765 (Electronic); 0947-6539 (Linking)

[9] Lan L, Yao Y, Ping J, Ying Y. Recent advances in nanomaterial-based biosensors for antibiotics detection. Biosensors and Bioelectronics. 2017;**91**:504-514. DOI: 10.1016/j.bios.2017.01.007. ISSN: 1873-4235 (Electronic); 0956-5663 (Linking)

[10] Liu Q, Ma C, Liu X-P, Wei Y-P, Mao C-J, Zhu J-J. A novel electrochemiluminescence biosensor for the detection of microRNAs based on a DNA functionalized nitrogen doped carbon quantum dots as signal enhancers. Biosensors and Bioelectronics. 2017;**92**:273-279. DOI: 10.1016/j.bios.2017.02.027. ISSN: 1873-4235 (Electronic); 0956-5663 (Linking)

[11] Baig MMF, Chen Y-C. Bright carbon dots as fluorescence sensing agents for bacteria and curcumin. Journal of Colloid and Interface Science. 2017;**501**:341-349. DOI: 10.1016/j.jcis.2017.04.045. ISSN: 1095-7103 (Electronic); 0021-9797 (Linking)

[12] Cheng C, Shi Y, Li M, Xing M, Wu Q. Carbon quantum dots from carbonized walnut shells: Structural evolution, fluorescence characteristics, and intracellular bioimaging. Materials Science and Engineering C: Materials for Biological Applications. 2017;**79**:473-480. DOI: 10.1016/j.msec.2017.05.094

[13] Guo J, Liu D, Filpponen I, Johansson L-S, Malho J-M, Quraishi S, Liebner F, Santos HA, Rojas OJ. Photoluminescent hybrids of cellulose nanocrystals and carbon quantum dots as cytocompatible probes for in vitro bioimaging. Biomacromolecules. 2017. DOI: 10.1021/acs.biomac.7b00306. ISSN: 1526-4602 (Electronic); 1525-7797 (Linking)

[14] Yang X, Yang W, Shen X, Chunyan S, Yang J, Piao M, Jia F, Gao G, Zhang L, Lin Q. One-step synthesis of photoluminescent carbon dots with excitation-independent emission for selective bioimaging and gene delivery. Journal of Colloid and Interface Science. 2017;**492**: 1-7. DOI: 10.1016/j.jcis.2016.12.057. ISSN: 1095-7103 (Electronic); 0021-9797 (Linking)

[15] Pan J, Zheng Z, Yang J, Wu Y, Lu F, Chen Y, Gao W. A novel and sensitive fluorescence sensor for glutathione detection by controlling the surface passivation degree of carbon quantum dots. Talanta. 2017;**166**:1-7. DOI: 10.1016/j.talanta.2017.01.033. ISSN: 1873-3573 (Electronic); 0039-9140 (Linking)

[16] Lee JJ, Yazan LS, Abdullah CAC. A review on current nanomaterials and their drug conjugate for targeted breast cancer treatment. International Journal of Nanomedicine. 2017;**12**:2373-2384. DOI: 10.2147/IJN.S127329. ISSN: 1178-2013 (Electronic); 1176-9114 (Linking)

[17] Chen Q, Wang Y, Wang Y, Zhang X, Duan D, Fan C. Nitrogen-doped carbon quantum dots/Ag_3PO_4 complex photocatalysts with enhanced visible light driven photocatalytic activity and stability. Journal of Colloid and Interface Science. 2017;**491**:238-245. DOI: 10.1016/j.jcis.2016.12.013. ISSN: 1095-7103 (Electronic); 0021-9797 (Linking)

[18] Ren Z, Liu X, Chu H, Yu H, Xu Y, Zheng W, Lei W, Chen P, Li J, Li C. Carbon quantum dots decorated $MoSe_2$ photocatalyst for Cr(VI) reduction in the UV-vis-NIR photon energy range. Journal of Colloid and Interface Science. 2017;**488**:190-195. DOI: 10.1016/j.jcis.2016.10.077. ISSN: 1095-7103 (Electronic); 0021-9797 (Linking)

[19] Ke J, Li X, Zhao Q, Liu B, Liu S, Wang S. Upconversion carbon quantum dots as visible light responsive component for efficient enhancement of photocatalytic performance. Journal of Colloid and Interface Science. 2017;**496**:425-433. DOI: 10.1016/j.jcis.2017.01.121. ISSN: 1095-7103 (Electronic); 0021-9797 (Linking)

[20] Di J, Xia J, Ji M, Li H, Xu H, Li H, Chen R. The synergistic role of carbon quantum dots for the improved photocatalytic performance of Bi_2MoO_6. Nanoscale. 2015;**7**(26): 11433-11443. ISSN: 2040-3372 (Electronic); 2040-3364 (Linking). DOI: 10.1039/c5nr01350j

[21] Di J, Xia J, Ge Y, Li H, Ji H, Xu H, Zhang Q, Li H, Li M. Novel visible-light-driven CQDs/Bi_2WO_6 hybrid materials with enhanced photocatalytic activity toward organic pollutants degradation and mechanism insight. Applied Catalysis. B, Environmental. 2015;**168**: 51-61. DOI: 10.1016/J.APCATB.2014.11.057. URL: http://www.sciencedirect.com/science/article/pii/S092633731400770X

[22] Liu X-Y, Chen H, Wang R, Shang Y, Zhang Q, Li W, Zhang G, Juan S, Dinh CT, Pelayo Garcia de Arquer F, Li J, Jiang J, Mi Q, Si R, Li X, Sun Y, Long Y-T, Tian H, Sargent EH, Ning Z. 0D-2D quantum dot: Metal dichalcogenide nanocomposite photocatalyst achieves efficient hydrogen generation. Advanced Materials. June 2017;**29**(22). ISSN: 1521-4095 (Electronic); 0935-9648 (Linking). DOI: 10.1002/adma.201605646

[23] Shi R, Li Z, Yu H, Lu S, Zhou C, Waterhouse GIN, Wu L-Z, Zhang T. Effect of nitrogen doping level on the performance of N-doped carbon quantum dot/TiO_2 composites for photocatalytic hydrogen evolution. ChemSusChem. 2017. DOI: 10.1002/cssc.201700943. ISSN: 1864-564X (Electronic); 1864-5631 (Linking)

[24] Martindale BCM, Joliat E, Bachmann C, Alberto R, Reisner E. Clean donor oxidation enhances the H_2 evolution activity of a carbon quantum dot-molecular catalyst photosystem. Angewandte Chemie (International Ed. in English). 2016;**55**(32):9402-9406. ISSN: 1521-3773 (Electronic); 1433-7851 (Linking). DOI: 10.1002/anie.201604355

[25] Xu X, Bao Z, Zhou G, Zeng H, Hu J. Enriching photoelectrons via three transition channels in amino-conjugated carbon quantum dots to boost photocatalytic hydrogen generation. ACS Applied Materials & Interfaces. 2016;**8(22)**:14118-14124. DOI: 10.1021/acsami.6b02961. ISSN: 1944-8252 (Electronic); 1944-8244 (Linking)

[26] Liu J, Yang L, Liu N, Han Y, Zhang X, Huang H, Lifshitz Y, Lee S-T, Zhong J, Kang Z. Metal-free efficient photocatalyst for stable visible water splitting via a two-electron pathway. Science. 2015;**347**(6225):970. URL: http://science.sciencemag.org/content/347/6225/970.abstract

[27] Li Q, Cui C, Meng H, Jiaguo Y. Visible-light photocatalytic hydrogen production activity of $ZnInS_4$ microspheres using carbon quantum dots and platinum as dual co-catalysts. Chemistry, an Asian Journal. 2014;**9**(7):1766-1770. DOI: 10.1002/asia.201402128. ISSN: 1861-471X (Electronic); 1861-471X (Linking)

[28] Shiral Fernando KA, Sahu S, Liu Y, Lewis WK, Guliants EA, Jafariyan A, Wang P, Bunker CE, Sun Y-P. Carbon quantum dots and applications in photocatalytic energy conversion. ACS Applied Materials & Interfaces. 2015;**7**(16):8363-8376. DOI: 10.1021/acsami.5b00448. ISSN: 1944-8252 (Electronic); 1944-8244 (Linking)

[29] Li H, Zhang X, MacFarlane DR. Carbon quantum dots/Cu_2O Heterostructures for solar-light-driven conversion of CO_2 to methanol. Advanced Energy Materials. 2015;**5**(5):1401077-n/a). DOI: 10.1002/aenm.201401077

[30] Sahu S, Liu Y, Wang P, Bunker CE, Shiral Fernando KA, Lewis WK, Guliants EA, Yang F, Wang J, Sun Y-P. Visible-light photoconversion of carbon dioxide into organic acids in an aqueous solution of carbon dots. Langmuir. 2014;**30**(28):8631-8636. ISSN: 1520-5827 (Electronic); 0743-7463 (Linking). DOI: 10.1021/la5010209

[31] Kou J, Lu C, Wang J, Chen Y, Xu Z, Varma RS. Selectivity enhancement in heterogeneous photocatalytic transformations. Chemical Reviews. 2017;**117**(3):1445-1514. DOI: 10.1021/acs.chemrev.6b00396. ISSN: 1520-6890 (Electronic); 0009-2665 (Linking)

[32] Mosconi D, Mazzier D, Silvestrini S, Privitera A, Marega C, Franco L, Moretto A. Synthesis and photochemical applications of processable polymers enclosing photoluminescent carbon quantum dots. ACS Nano. 2015;**9**(4):4156-4164. DOI: 10.1021/acsnano.5b00319. ISSN: 1936-086X (Electronic); 1936-0851 (Linking)

[33] Tomita O, Ohtani B, Abe R. Highly selective phenol production from benzene on a platinum-loaded tungsten oxide photocatalyst with water and molecular oxygen: Selective oxidation of water by holes for generating hydroxyl radical as the predominant source of the hydroxyl group. Catalysis Science & Technology. 2014;**4**(11):3850-3860. DOI: 10.1039/C4CY00445K

[34] Yu H, Shi R, Zhao Y, Waterhouse GIN, Wu L-Z, Tung C-H, Zhang T. Smart utilization of carbon dots in semiconductor photocatalysis. Advanced Materials. 2016;**28**(43): 9454-9477. ISSN: 1521-4095 (Electronic); 0935-9648 (Linking). DOI: 10.1002/adma.2016 02581

[35] Gaya UI, Abdullah AH. Heterogeneous photocatalytic degradation of organic contaminants over titanium dioxide: A review of fundamentals, progress and problems. Journal of Photochemistry and Photobiology. 2008;**9**(1):1-12

[36] Konstantinou IK, Albanis TA. (TiO_2)-assisted photocatalytic degradation of azo dyes in aqueous solution: Kinetic and mechanistic investigations: A review. Applied Catalysis. B, Environmental. 2004;**49**(1):1-14

[37] Houas A, Lachheb H, Ksibi M, Elaloui E, Guillard C, Herrmann J-M. Photocatalytic degradation pathway of methylene blue in water. Applied Catalysis. B, Environmental. 2001;**31**(2):145-157

[38] Herrmann J-M. Heterogeneous photocatalysis: Fundamentals and applications to the removal of various types of aqueous pollutants. Catalysis Today. 1999;**53**(1):115-129. DOI: 10.1016/S0920-5861(99)00107-8 URL: http://www.sciencedirect.com/science/article/pii/S0920586199001078

[39] Anders, Hagfeldt, Michael, Graetzel. Light-induced redox reactions in nanocrystalline systems. Chemical Reviews. 1995;**95**(1):49-68. DOI: 10.1021/cr00033a003

[40] Hoffmann MR, Martin ST, Choi W, Bahnemann DW. Environmental applications of semiconductor photocatalysis. Chemical Reviews. 1995;**95**(1):69-96. DOI: 10.1021/cr00033a004

[41] Fox MA, Dulay MT. Heterogeneous photocatalysis. Chemical Reviews. 1993;**93**(1):341-357. DOI: 10.1021/cr00017a016

[42] Linsebigler AL, Guangquan L, Yates JT. Photocatalysis on TiO_2 surfaces: Principles, mechanisms, and selected results. Chemical Reviews. 1995;**95**(3):735-758. DOI: 10.1021/cr00035a013

[43] Habisreutinger SN, Schmidt-Mende L, Stolarczyk JK. Photocatalytic reduction of CO_2 on TiO_2 and other semiconductors. Angewandte Chemie (International Ed. in English). July 2013;**52**(29):7372-7408. DOI: 10.1002/anie.201207199. ISSN: 1521-3773 (Electronic); 1433-7851 (Linking)

[44] Zhou H, Yongquan Q, Zeid T, Duan X. Towards highly efficient photocatalysts using semiconductor nanoarchitectures. Energy & Environmental Science. 2012;**5**(5):6732-6743

[45] Ni M, Leung MKH, Leung DYC, Sumathy K. A review and recent developments in photocatalytic water-splitting using TiO_2 for hydrogen production. Renewable and Sustainable Energy Reviews. 2007;**11**(3):401-425. ISSN: 1364-0321

[46] Brown TE, Brown TL, LeMay HEH, Bursten BE, Murphy C, Woodward P. Chemistry: The Central Science. Vol. 13, illustrated. Pearson; 2015. ISBN: 9780321910417. URL: https://books.google.com.ng/books?id=fhzpngEACAAJ

[47] Kisch H. Semiconductor photocatalysis—Mechanistic and synthetic aspects. Angewandte Chemie (International Ed. in English). 2013;**52**(3):812-847. DOI: 10.1002/anie.201201200

[48] Hakki A, Schneider J, Bahnemann D. Understanding the Chemistry of Photocatalytic Processes. Photocatalysis: Fundamentals and Perspectives. The Royal Society of Chemistry; 2016. Chapter 2. pp. 29-50. ISBN: 978-1-78262-041-9. DOI: 10.1039/9781782622338-00029

[49] Ohtani B. Revisiting the fundamental physical chemistry in heterogeneous photocatalysis: Its thermodynamics and kinetics. Physical Chemistry Chemical Physics. 2014;**16**(5):1788-1797. DOI: 10.1039/C3CP53653J

[50] Herrmann J-M, Guillard C, Pichat P. Heterogeneous photocatalysis: An emerging technology for water treatment. Catalysis Today. 1993;**17**(1–2):7-20

[51] Wu W, Jiang C, Roy VAL. Recent progress in magnetic iron oxide-semiconductor composite nanomaterials as promising photocatalysts. Nanoscale. 2015;**7**(1):38-58. DOI: 10.1039/C4NR04244A

[52] Rothenberger G, Comte P, Grätzel M. A contribution to the optical design of dye-sensitized nanocrystalline solar cells. Solar Energy Materials & Solar Cells. 1999;**58**(3):321-336. DOI: 10.1016/S0927-0248(99)00015-X. URL: http://www.sciencedirect.com/science/article/pii/S092702489900015X

[53] Maeda K, Sahara G, Eguchi M, Ishitani O. Hybrids of a ruthenium(II) polypyridyl complex and a metal oxide nanosheet for dye-sensitized hydrogen evolution with visible light: Effects of the energy structure on photocatalytic activity. ACS Catalysis. 2015;**5**(3):1700-1707. DOI: 10.1021/acscatal.5b00040

[54] Ozawa H, Honda S, Katano D, Sugiura T, Arakawa H. Novel ruthenium sensitizers with a dianionic tridentate ligand for dye-sensitized solar cells: The relationship between the solar cell performances and the electron-withdrawing ability of substituents on the ligand. Dalton Transactions. 2014;**43**:8026-8036. DOI: 10.1039/C3DT52873A

[55] Ardo S, Meyer GJ. Photodriven heterogeneous charge transfer with transition-metal compounds anchored to TiO_2 semiconductor surfaces. Chemical Society Reviews. 2009;**38**:115-164. DOI: 10.1039/B804321N

[56] Zhang X, Yu L, Wu J, Jing J, Fang B. Improved performance of CdSe/CdS/PbS co-sensitized solar cell with double-layered TiO_2 films as photoanode. Optics Communication. 2017;**395**:117-121. DOI: 10.1016/j.optcom.2016.05.026. URL: http://www.sciencedirect.com/science/article/pii/S0030401816303789

[57] Sakai T, Mersch D, Reisner E. Photocatalytic hydrogen evolution with a hydrogenase in a mediator-free system under high levels of oxygen. Angewandte Chemie, International Edition. 2013;**52**(47):12313-12316. DOI: 10.1002/anie.201306214

[58] Ye K-H, Wang Z, Jiuwang G, Xiao S, Yuan Y, Zhu Y, Zhang Y, Mai W, Yang S. Carbon quantum dots as a visible light sensitizer to significantly increase the solar water splitting

performance of bismuth vanadate photoanodes. Energy & Environmental Science. 2017; **10**(3):772-779. DOI: 10.1039/C6EE03442J

[59] Lim SY, Shen W, Gao Z. Carbon quantum dots and their applications. Chemical Society Reviews. 2015;**44**(1):362-381

[60] Martindale BCM, Hutton GAM, Caputo CA, Reisner E. Solar hydrogen production using carbon quantum dots and a molecular nickel catalyst. Journal of the American Chemical Society. 2015;**137**(18):6018-6025. DOI: 10.1021/jacs.5b01650

[61] Jin Q, Gubu A, Chen X, Tang X. A photochemical avenue to photoluminescent N-dots and their upconversion cell imaging. Scientific Reports. 2017;**7**(1):1793. DOI: 10.1038/s41598-017-01663-x

[62] Xie R, Zhang L, Xu H, Zhong Y, Sui X, Mao Z. Construction of up-converting fluorescent carbon quantum dots/Bi_2OTiO_3 2 composites with enhanced photocatalytic properties under visible light. Chemical Engineering Journal. 2017;**310**(Part 1):79-90. DOI: 10.1016/j.cej.2016.10.089. URL: http://www.sciencedirect.com/science/article/pii/S1385894716314991

[63] Zhang X, Huang H, Liu J, Yang L, Kang Z. Carbon quantum dots serving as spectral converters through broadband upconversion of near-infrared photons for photoelectrochemical hydrogen generation. Journal of Materials Chemistry A. 2013;**1**(38):11529-11533. DOI: 10.1039/C3TA12568H

[64] Wang R, Lu K-Q, Tang Z-R, Yi-Jun X. Recent progress in carbon quantum dots: Synthesis, properties and applications in photocatalysis. Journal of Materials Chemistry C. 2017; **5**(8):3717-3734. DOI: 10.1039/C6TA08660H

[65] Wang X, Cao L, Lu F, Meziani MJ, Li H, Qi G, Zhou B, Harruff BA, Kermarrec F, Sun Y-P. Photoinduced electron transfers with carbon dots. Chemical Communications (Camb). 2009;**25**:3774-3776. DOI: 10.1039/B906252A. URL: http://www.ncbi.nlm.nih.gov/pmc/articles/PMC2767382/

[66] Yu H, Zhang H, Huang H, Yang L, Li H, Ming H, Kang Z. ZnO/carbon quantum dots nanocomposites: One-step fabrication and superior photocatalytic ability for toxic gas degradation under visible light at room temperature. New Journal of Chemistry. 2012; **36**(4):1031-1035

[67] Chen P, Wang F, Chen Z-F, Zhang Q, Yuehan S, Shen L, Yao K, Yang L, Cai Z, Lv W, Liu G. Study on the photocatalytic mechanism and detoxicity of gemfibrozil by a sunlight-driven TiO_2/carbon dots photocatalyst: The significant roles of reactive oxygen species. Applied Catalysis. B, Environmental. 2017;**204**(Supplement C):250-259. DOI: 10.1016/j.apcatb.2016.11.040. URL: http://www.sciencedirect.com/science/article/pii/S0926337316309079

[68] Virca CN, Winter HM, Goforth AM, Mackiewicz MR, McCormick TM. Photocatalytic water reduction using a polymer coated carbon quantum dot sensitizer and a nickel nanoparticle catalyst. Nanotechnology. 2017;**28**(19):195402. URL: http://stacks.iop.org/0957-4484/28/i=19/a=195402

[69] Prasannan A, Imae T. One-pot synthesis of fluorescent carbon dots from orange waste peels. Industrial and Engineering Chemistry Research. 2013;**52**(44):15673-15678

[70] Zhang H, Huang H, Ming H, Li H, Zhang L, Yang L, Kang Z. Carbon quantum dots/ Ag_3PO_4 complex photocatalysts with enhanced photocatalytic activity and stability under visible light. Journal of Materials Chemistry. 2012;**22**(21):10501-10506

[71] Yu X, Liu J, Yu Y, Zuo S, Li B. Preparation and visible light photocatalytic activity of carbon quantum dots/TiO_2 nanosheet composites. Carbon. 2014;**68**:718-724

[72] Wang F, Chen P, Feng Y, Xie Z, Yang L, Yuehan S, Zhang Q, Wang Y, Yao K, Lv W, Liu G. Facile synthesis of N-doped carbon dots/g-C_3N_4 photocatalyst with enhanced visible-light photocatalytic activity for the degradation of indomethacin. Applied Catalysis. B, Environmental. 2017;**207**(Supplement C):103-113. DOI: 10.1016/j.apcatb.2017.02.024. URL: http://www.sciencedirect.com/science/article/pii/S0926337317301285

[73] Boxi SS, Paria S. Visible light induced enhanced photocatalytic degradation of organic pollutants in aqueous media using Ag doped hollow TiO_2 nanospheres. RSC Advances. 2015;**5**(47):37657-37668

[74] Nakajima A, Kobayashi T, Isobe T, Matsushita S. Preparation and visible-light photocatalytic activity of Au-supported porous CeO_2 spherical particles using templating. Materials Letters. 2011;**65**(19):3051-3054. DOI: 10.1016/J.MATLET.2011.06.051. URL: http://www.sciencedirect.com/science/article/pii/S0167577X1100694X

[75] Xiong D, Chang H, Zhang Q, Tian S, Liu B, Zhao X. Preparation and characterization of $CuCrO_2$/TiO_2 heterostructure photocatalyst with enhanced photocatalytic activity. Applied Surface Science. 2015;**347**:747-754

[76] Xiong X, Chen H, Yiming X. Improved photocatalytic activity of TiO_2 on the addition of $CuWO_4$. Journal of Physical Chemistry C. Mar. 2015;**119**(11):5946-5953. DOI: 10.1021/ jp510974f

[77] Wang W, Wang J, Wang Z, Wei X, Liu L, Ren Q, Gao W, Liang Y, Shi H. p-n junction CuO/ $BiVO_4$ heterogeneous nanostructures: Synthesis and highly efficient visible-light photocatalytic performance. Dalton Transactions. 2014;**43**(18):6735-6743

[78] Wang M, Shen S, Li L, Tang Z, Yang J. Effects of sacrificial reagents on photocatalytic hydrogen evolution over different photocatalysts. Journal of Materials Science. May 2017; **52**(9):5155-5164. ISSN: 1573-4803. DOI: 10.1007/s10853-017-0752-z

[79] Li X, Low J, Jiaguo Y. Photocatalytic hydrogen generation. Photocatalysis: Applications. In: Dionysiou DD, Puma GL, JinhuaYe JS, Bahnemann D, editors. Energy and Environment Series. The Royal Society of Chemistry; 2016. Chapter 10. pp. 255-302. ISBN: 978-1-78262-709-8. DOI: 10.1039/9781782627104-00255

[80] Park H, Kim H-i, Moon G-h, Choi W. Photoinduced charge transfer processes in solar photocatalysis based on modified TiO_2. Energy & Environmental Science. 2016;**9**(2):411-433. DOI: 10.1039/C5EE02575C

[81] Xiang Q, Cheng B, Jiaguo Y. Graphene-based Photocatalysts for solar-fuel generation. Angewandte Chemie, International Edition. 2015;**54**(39):11350-11366

[82] Rawalekar S, Mokari T. Rational design of hybrid nanostructures for advanced photocatalysis. Advanced Energy Materials. 2013;**3**(1):12-27. DOI: 10.1002/aenm.201200511

[83] Zhang L-W, Hong-Bo F, Zhu Y-F. Efficient TiO_2 photocatalysts from surface hybridization of TiO_2 particles with graphite-like carbon. Advanced Functional Materials. 2008;**18**(15): 2180-2189. DOI: 10.1002/adfm.200701478

[84] Di J, Xia J, Ji M, Wang B, Yin S, Xu H, Chen Z, Li H. Carbon quantum dots induced ultrasmall BiOI nanosheets with assembled hollow structures for broad spectrum photocatalytic activity and mechanism insight. Langmuir. 2016;**32**(8):2075-2084. DOI: 10.1021/acs.langmuir.5b04308

[85] Sun M, Ma X, Chen X, Sun Y, Cui X, Lin Y. A nanocomposite of carbon quantum dots and TiO_2 nanotube arrays: Enhancing photoelectrochemical and photocatalytic properties. RSC Advances. 2014;**4**(3):1120-1127

[86] Li H, He X, Kang Z, Huang H, Yang L, Liu J, Lian S, Tsang CHA, Yang X, Lee S-T. Water-soluble fluorescent carbon quantum dots and photocatalyst design. Angewandte Chemie, International Edition. 2010;**49**(26):4430-4434

[87] Li J, Ma Y, Ye Z, Zhou M, Wang H, Ma C, Wang D, Huo P, Yan Y. Fast electron transfer and enhanced visible light photocatalytic activity using multi-dimensional components of carbon quantum dots @ 3D daisy-like In_2S_3/single-wall carbon nanotubes. Applied Catalysis. B, Environmental. 2017;**204**:224-238. DOI: 10.1016/J.APCATB.2016.11.021. URL: http://www.sciencedirect.com/science/article/pii/S0926337316308797

[88] Chai N-N, Wang H-X, Hu C-X, Wang Q, Zhang H-L. Wellcontrolled layer-by-layer assembly of carbon dot/CdS heterojunctions for efficient visiblelight- driven photocatalysis. Journal of Materials Chemistry A. 2015;**3**(32):16613-16620. DOI: 10.1039/C5TA03649F

[89] Huo P, Guan J, Zhou M, Ma C, Liu X, Yan Y, Yuan S. Carbon quantum dots modified CdSe loaded reduced graphene oxide for enhancing photocatalytic activity. Journal of Industrial and Engineering Chemistry. 2017;**50**:147-154. DOI: 10.1016/J.JIEC.2017.02.008. URL: http://www.sciencedirect.com/science/article/pii/S1226086X17300709

[90] Makama AB, Salmiaton A, Saion EB, Choong TSY, Abdullah N. Microwave-assisted synthesis of porous ZnO/SnS_2 heterojunction and its enhanced photoactivity for water purification. Journal of Nanomaterials. 2015;**2015**:13. DOI: 10.1155/2015/108297

[91] Xie X, Kretschmer K, Wang G. Advances in graphene-based semiconductor photocatalysts for solar energy conversion: Fundamentals and materials engineering. Nanoscale. 2015; **7**(32):13278-13292. DOI: 10.1039/C5NR03338A

[92] Han C, Yang M-Q, Weng B, Yi-Jun X. Improving the photocatalytic activity and anti-photocorrosion of semiconductor ZnO by coupling with versatile carbon. Physical Chemistry Chemical Physics. 2014;**16**(32):16891-16903. DOI: 10.1039/C4CP02189D

[93] Arul V, Edison TNJI, Lee YR, Sethuraman MG. Biological and catalytic applications of green synthesized fluorescent N-doped carbon dots using *Hylocereus undatus*. Journal of Photochemistry and Photobiology B. 2017;**168**:142-148. DOI: 10.1016/j.jphotobiol.2017.02.007

[94] Chen J, Che H, Huang K, Liu C, Shi W. Fabrication of a ternary plasmonic photocatalyst CQDs/Ag/Ag_2O to harness charge flow for photocatalytic elimination of pollutants. Applied Catalysis B. 2016;**192**:134-144. DOI: 10.1016/J.APCATB.2016.03.056. URL: http://www.sciencedirect.com/science/article/pii/S0926337316302387

[95] Xia J, Di J, Li H, Xu H, Li H, Guo S. Ionic liquid-induced strategy for carbon quantum dots/BiOX (X = Br, Cl) hybrid nanosheets with superior visible light-driven photocatalysis. Applied Catalysis. B, Environmental. 2016;**181**:260-269. DOI: 10.1016/J.APCATB.2015.07.035. URL: http://www.sciencedirect.com/science/article/pii/S0926337315300540

[96] Byong Yong Yu And Seung-Yeop Kwak. Carbon quantum dots embedded with mesoporous hematite nanospheres as efficient visible light-active photocatalysts. Journal of Materials Chemistry 2012;**22**(17):8345-8353

[97] Li H, Liu R, Yang L, Huang H, Yu H, Ming H, Lian S, Lee S-T, Kang Z. Carbon quantum dots/Cu_2O composites with protruding nanostructures and their highly efficient (near) infrared photocatalytic behavior. Journal of Materials Chemistry. 2012;**22**(34):17470-17475

[98] Di J, Xia J, Ji M, Xu L, Yin S, Zhang Q, Chen Z, Li H. Carbon quantum dots in situ coupling to bismuth oxyiodide via reactable ionic liquid with enhanced photocatalytic molecular oxygen activation performance. Carbon. 2016;**98**(Supplement C):613-623. DOI: 10.1016/j.carbon.2015.11.015. URL: http://www.sciencedirect.com/science/article/pii/S0008622315304243

Chapter 8

Visible Light–Driven Hydrogen Production by Carbon based Polymeric Materials

Palas Baran Pati, SU-IL In and Haining Tian

Additional information is available at the end of the chapter

http://dx.doi.org/10.5772/intechopen.70444

Abstract

Converting solar energy into storable solar fuels such as H_2 from earth abundant source—water—is a nice approach to find the solution of energy crisis and environmental protection. There are two half reactions; first, water oxidation into oxygen and proton and followed by proton reduction led to H_2 evolution from water. After two decades of continuous attempts, there have been several efficient water oxidation photocatalysts introduced, whereas the proton reduction photocatalyst were relatively less explored. Major portion of reported photocatalysts for proton reduction are mainly derived from either noble metals or precious metals. Carbon-based organic photocatalysts have become attractive recently. These organic materials have several advantages like light weight, cheap, well-defined structure-property relationship and the most attractive one is better batch to batch reproducibility. Here, the reported organic photocatalysts and their performance are summarized which in fact help others to get an idea about ongoing progress in this area of research and to understand the basic designing principle for efficient photocatalysts for fuel production.

Keywords: solar fuels, photochemical hydrogen generation, proton reduction, photocatalysts, carbon nitride, conjugated polymers, polymer dots, porous materials

1. Introduction

Due to the limited storage of fossil fuels and the environmental hazards from burning of fossil fuels, there is urgent need for renewable energy sources. Harvesting solar energy is an ideal alternative solution. Silicon solar cell is already proven as an efficient technology for converting solar energy into electrical energy. In past two decades, enormous attempts have been done to replace silicon solar cells by organic carbon-based compound such as dyes and polymers for developing efficient solar energy harvesting cells [1, 2]. But all these solar cells can convert solar energy into electrical energy. From the total consumption of energy in our society, we can see that only around 20% of the energy consumed is electrical energy and the rest is used as fuel form. So, storing solar energy into fuels is another important solution to energy crisis.

The light-induced water splitting into chemical fuels, oxygen and hydrogen, by sunlight represents an unlimited source of energy from naturally abundant resources. To carry out the process, it requires a photocatalyst and the key half reactions of water splitting are the oxidation of water to form oxygen and protons and the reduction of the protons to form hydrogen. The standard reduction potentials of these processes (pH = 0) are 1.23 V and 0 V vs. Normal Hydrogen Electrode (NHE), respectively, and the catalyst has to utilise photons of sufficient energy to generate at least this potential in order to split water, in practise a significant over potential usually results in a potential of at least 2V being required [3]. Groundbreaking work by Fujishima and Honda in 1972 showed that hydrogen generation *via* water splitting is possible using photocatalysts based on oxide semiconductors capable of adsorbing ultraviolet light energy [4]. For efficient hydrogen generation, development of efficient proton reduction catalyst is very essential. Enormous attempts have been done during the last few years to develop efficient proton reduction catalyst. Molecular catalyst-photosensitizer assemblies and inorganic material photocatalysts have been intensively developed [5]. On the other hand, organic photocatalysts have also attracted intense interests from scientists due to their facile structure modification, less environmental footprints, tunable light absorption and promising photocatalytic reactivity.

Conjugated materials can absorb visible light because of their delocalized π-system. These properties have led to applications in organic electronics and organic photonics [2]. In early 1985, Yanagida et al. have used linear poly(p-phenylene) for hydrogen generation which is known as the first report on carbon-based material for photochemical hydrogen generation [6]. This material only showed very low activity (apparent quantum yield (AQY) = 0.006%) and worked under ultraviolet (UV)-light (λ >366 nm). Another disadvantage was the insolubility, which delimits the further advancement on these materials. After two decades, carbon nitride has established itself as an efficient proton reduction catalyst [7]. Carbon nitride (g-C_3N_4) shows good efficiency on proton reduction in the presence of some co-catalysts such as Ni and Ni-P [8]. In the presence of co-catalyst, carbon nitride even can oxidize water [9]. But in the absence of co-catalysts, carbon nitride showed unsatisfactory performance for proton reduction. Due to the fixed structure of carbon nitride, there are few chances to tune the properties mainly light harvesting ability, which delimits the efficiency of the materials. Due to the high activity of carbon nitride and the restriction towards further improvement, scientists started to look and use conjugated materials for solar light-driven hydrogen production. The major advantage of the carbon-based organic semiconductor is that easily synthetic and purifying process and better batch to batch reproducibility. Well-established structure-property relationship helps to design and modify the structure to improve the performance of the materials. Very recently, various pure organic semiconducting materials such as poly(azomethine)s [10], covalent organic framework [11], phenyl-triazine oligomers [12], microporous organic nanorods [13], heptazine networks [14] and polybenzothiadiazoles [15] have been successfully applied in the photocatalytic hydrogen production. These organic materials work as heterogeneous catalyst which associated with poor dispersibility in aqueous medium. The difficult dispersibility results in the low efficiency and stability of these materials. So, there is chance and need for improvement and betterment in this field to get highly efficient organic photocatalysts for photochemical hydrogenation.

Herein, in this chapter, we summarized and discussed the facile synthesis and latest developments of carbon-based photo-active organic materials usable for photochemical hydrogen generation. We will discuss the working principle and the thermodynamics behind the process for showing proton reduction ability to generate hydrogen. The discussion of available organic materials will be subdivided into following types: (a) covalent organic frameworks (COFs), (b) covalent triazine frameworks (CTFs), (c) carbon nitride-based polymers, (d) conjugated microporous polymers (CMP), (e) conjugated linear polymers and very recently introduced (f) polymer dots (Pdots). Our discussion particularly focuses on the modification of synthesis, structure and properties, aiming at better understanding of the intimate structure-performance relationship and further enhancement of photocatalytic activities. We believe that this chapter will be helpful for the readers to understand the working principle and structure-property relationship to devise a high-efficient photocatalyst. It is very timely on the aspect of renewable energy source finding.

2. Thermodynamics and driving force for proton reduction hydrogen generation process

Figure 1 shows the thermodynamics of water splitting (**Figure 1(a)**, at pH = 0) and photochemical hydrogen generation steps from water. Both the half reaction water oxidation and proton reduction are pH dependent. For the efficient water oxidation, the band gap of the conjugated material should be higher than 1.23 eV with well-stabilized oxidized state of the material

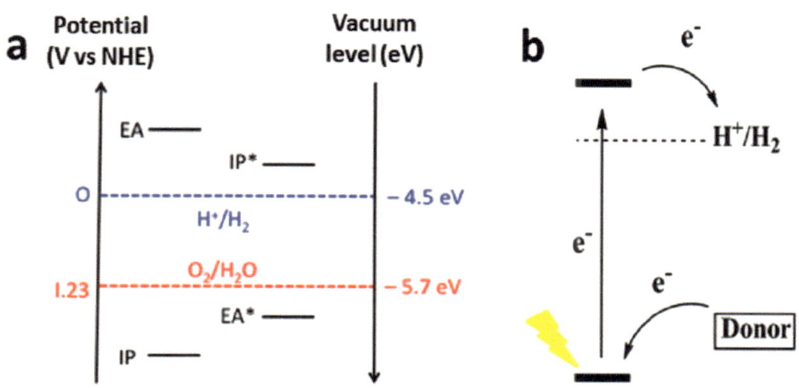

Figure 1. (a) Energy diagram of water splitting reaction. (b) Photochemical hydrogen generation energy diagram in the presence of a sacrificial electron donor.

to donate photogenerated hole for water oxidation to oxygen and proton. The excited state should be higher in energy to donate photoexcited electron to produce hydrogen upon reduction of proton. In case of photochemical hydrogen production (**Figure 1(b)**), the photoexcited material gets reduced by oxidizing sacrificial electron donor and the reduced material carries out the reduction of proton to produce hydrogen. There are many sacrificial donors used for the reaction; the most popular ones are triethanolamine (TEOA) and sodium ascorbate.

3. Materials

3.1. Covalent organic framework

Covalent organic frameworks (COFs) are highly ordered two- or three-dimensional crystalline polymers [16, 17]. These materials specially attract the interest in the area of photochemical hydrogen generation due to high surface area and tuneable pore size, which are essential criteria for heterogeneous catalysis [18]. By varying the molecular synthons, the optical and electrical properties of the resulted COFs can be nicely tuned. Stegbauer et al. enriched the library on this area by synthesizing and characterizing a series of COFs and further they successfully applied these materials for photochemical hydrogen generation in the presence of co-catalyst and a sacrificial electron donor [11]. Acid-catalysed Schiff base reaction resulted a two-dimensional (2D) mesoporous hydrazine-based COFs (TFTP-COF) with honeycomb-type network (**Figure 2**). It showed a high surface area of 1603 $m^2\ g^{-1}$ with a pore size of 3.8 nm. These polymers with a band gap of 2.8 eV show light absorption ability in the visible range. It shows high hydrogen evolution performance (230 $\mu mol\ h^{-1}\ g^{-1}$) when Pt nanoparticle and sodium ascorbate were used as co-catalyst and sacrificial electron donor, respectively, under visible light (>420 nm) irradiation. Indeed, it showed higher hydrogen evolution rate (1970 $\mu mol\ h^{-1}\ g^{-1}$) with a AQY of 2.2% when triethanolamine (TEOA) was used instead of sodium ascorbate as sacrificial electron donor.

Again Vyas et al. further synthesize a series of 2D azine-linked COFs with varying amount of nitrogen by solvothermal method from hydrazine and triphenylarylaldehydes [19]. The high BET-surface area (1537 $m^2\ g^{-1}$) suggests that the COFs are highly porous. All COFs absorb in the UV region and the blue part of the visible region with an absorption edge around 465–475 nm which led to an optical band gap of 2.6–2.7 eV (**Figure 3**). The **N_3-COF** showed hydrogen evolution rate of 1703 $\mu mol\ h^{-1}\ g^{-1}$ when TEOA used as sacrificial electron donor and hexachloroplatinic acid for in situ generation of Pt-nanoparticle.

The above study suggests that with the increase of N containing in the central core of COFs the photocatalytic activity is increased so further they tried to introduced N atom in the peripheral unit of COF also [20]. They introduced pyridine containing azine-linked COFs by solvothermal process between **PTP-CHO** and hydrazine hydrate (**Figure 4**). Photocatalytic hydrogen evolution rate was tested in the presence of TEOA as electron donor and hexachloroplatinic acid as in situ formation for Pt-nanoparticle used as co-catalyst. The hydrogen evolution rate is lower for **PTP-COF** (83.83 $\mu mol\ h^{-1}\ g^{-1}$) than **N_3-COF** (1703 $\mu mol\ h^{-1}\ g^{-1}$).

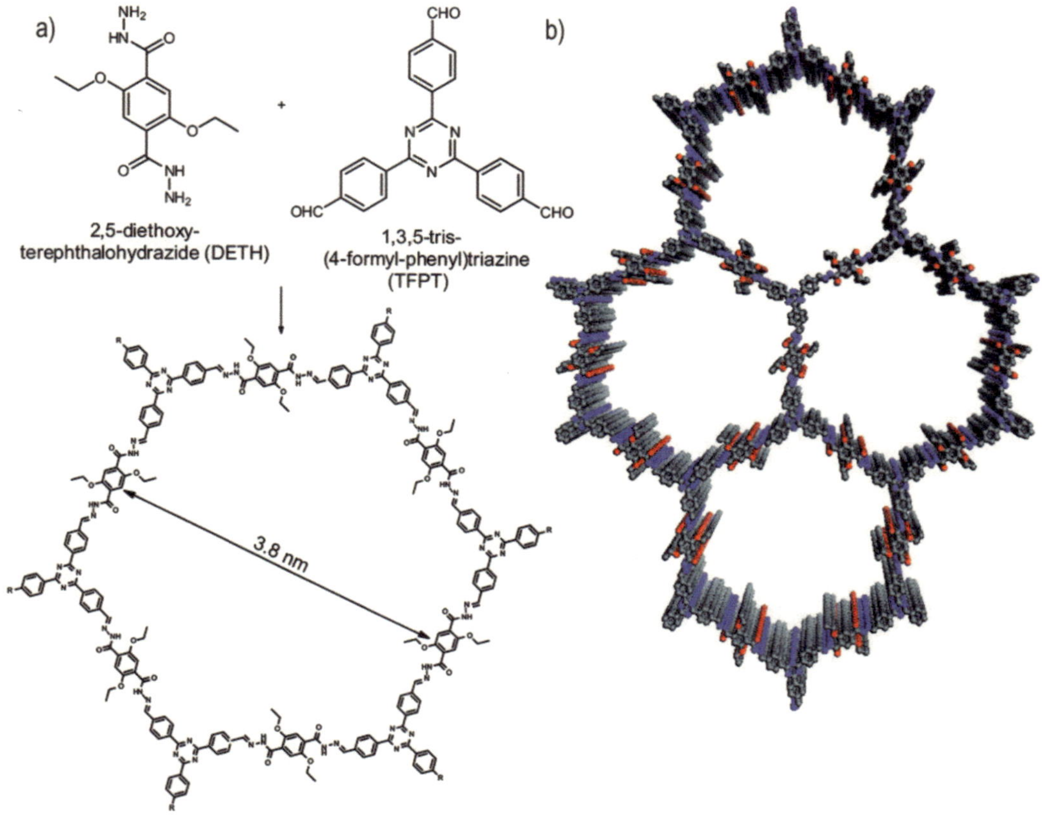

Figure 2. Acetic acid-catalysed hydrazone formation furnishes a mesoporous 2D network with a honeycomb type in plane structure. (a) Scheme showing the condensation of the two monomers to form the TFPT-COF. (b) TFPT-COF with a cofacial orientation of the aromatic building blocks, constituting a close-to eclipsed primitive hexagonal lattice (grey: carbon, blue: nitrogen, red: oxygen). Reproduced with permission from RSC; see Ref. [11].

3.2. Covalent triazine-based frameworks (CTFs)

It is believed that due to incorporation of electronegative nitrogen centre the H_2 evolution can be strongly enhanced. On this aspect, triazine or heptazine framework in the organic polymeric system can be beneficial for photochemical hydrogen production. CTFs are typically prepared by ionothermal trimerization of aromatic nitriles such as 1,4-dicyanobenzene (DCB), 1,3,5-tris(4-cyanophenyl)benzene (TCPB) and 2,6-dicyanopyridine (DCP) process in the presence of $ZnCl_2$ which acts both as catalyst and solvent [21]. These systems are typically microporous structure with high surface area. Typically, CTFs are black in colour with very low band gap (~1 eV) which in fact is insufficient for both photochemical water splitting and generated proton reduction. Schwinghammer et al. synthesized yellow phenyl-triazine oligomers by reducing carbonization effect through lowering temperature (300 °C) and prolonging reaction time (150 h) [12]. These synthesized CTFs showed the band gap in the range of 3.2–3.3 eV. In the presence of Pt as co-catalyst and TEOA as sacrificial donor, the most active CTF sample showed a hydrogen evolution rate of 1076 (±278) μmol h^{-1} g^{-1}. Although the hydrogen

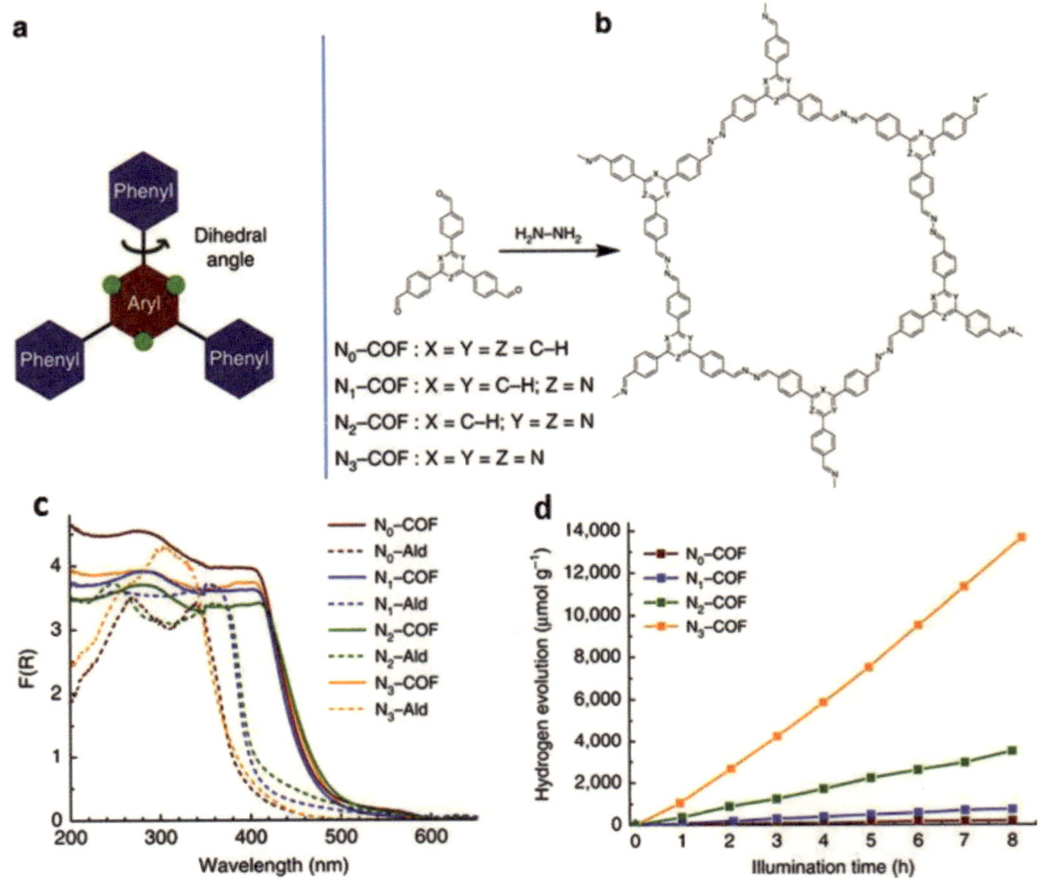

Figure 3. Design (a) and synthesis (b) of the **N_x-COFs**. Absorption spectra of precursor aldehydes **N_x-Alds** in dichloromethane at 22°C (c). Hydrogen production monitored over 8 h using **N_x-COFs** as photocatalyst in the presence of triethanolamine as sacrificial electron donor (d). Reproduced with permission from Nature Publication Group; see Ref. [19].

evolution rate (121 μmol h^{-1} g^{-1}) is low but most active CTF is active in photochemical hydrogen production in the absence of co-catalyst. The apparent quantum yield (AQY) of the highest active CTF is determined to be 5.5% at 400 (±20) nm. On an intension of efficient CTFs by manipulating structural and optical properties, Wu et al. introduced the DCB-based **CTF-T1** by incorporating methyl group in monomer through trifluoromethanesulfonic acid-catalysed reaction. The optical band gap of devised CTF was determined to be 2.9 eV. The CTF showed water splitting to produce H_2 in the presence of sacrificial electron donor. The AQY of hydrogen evolving reaction was calculated to 2.4% in 400–440 nm **(Figure 5)**.

3.3. Carbon nitride-based polymers

Since 2009, after successful demonstration in water splitting reaction, carbon nitride analogues become matter of interest. Commonly, it is believed that two different analogues, namely *s*-triazine

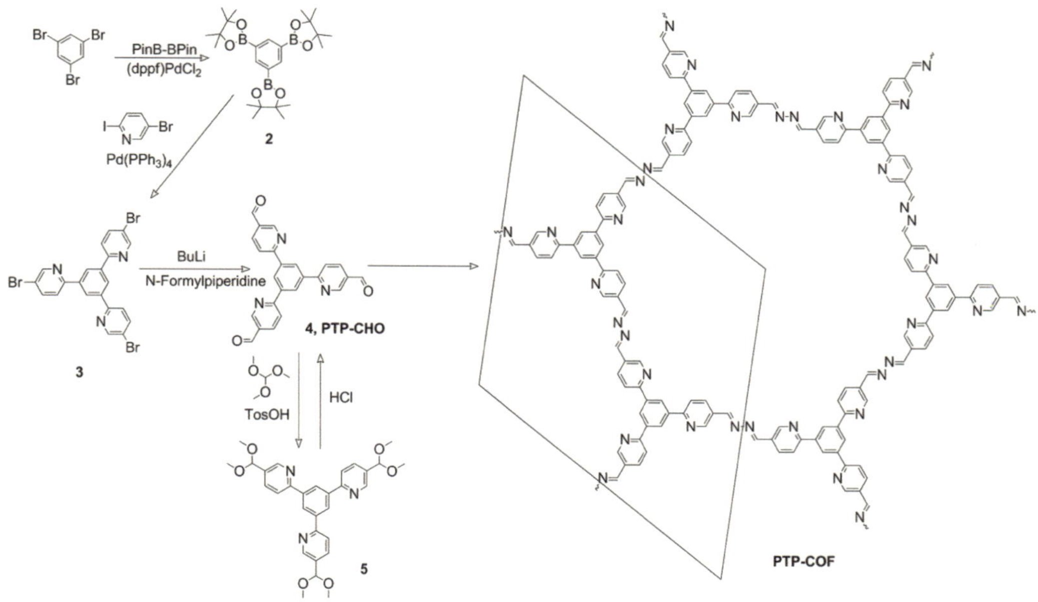

Figure 4. Synthesis of precursor **PTP-CHO** and **PTP-COF**. The in-plane unit cell is indicated on the right. Reproduced with permission from RSC; see Ref. [20].

and tri-*s*-triazine (heptazine), are co-existed to separately construct the ideal polymeric carbon nitride units [22–25]. Being energetically higher s-triazine is thermodynamically unfavourable to synthesize carbon nitride with pure s-triazine unit. In fact, poly(triazine imide) (PTI) and melon, which belong to typical s-triazine and tri-*s*-triazine structure, respectively, are always experimentally synthesized to obtain the binary carbon nitride polymers with 2D electronic structures [22–29]. PTI always features high level of crystallinity, whereas most of the melons are amorphous or semi-crystalline in structure. Besides, quite different optical properties were also generated due to the different degrees of conjugation. In this case, these two counterparts showed different photocatalytic performances for photocatalytic H_2 evolution. Ham et al. first reported that PTI/Li^+Cl^- exhibited a steady H_2 evolution rate under visible light irradiation in the presence

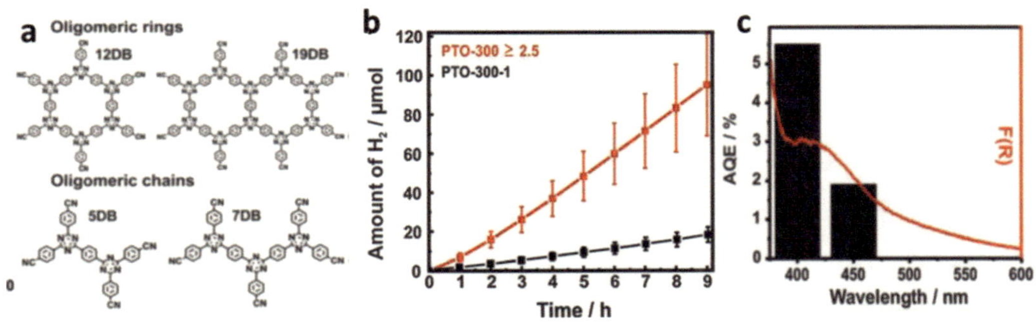

Figure 5. (a) Possible ring- and chain-like oligomers, (b) average hydrogen evolution curve and (c) quantum yield with a function of wavelength. Reproduced with permission from RSC; see Ref. [12].

of sacrificial agents and co-catalyst (Pt, 5 wt.%), whereas it showed low activity in comparison to that of Pt/g-C_3N_4 which may be assigned to the weak optical absorption in the visible light region [30]. Schwinghammer et al. modify the PTI/Li^+Cl^- by using 4-amino-2, 6-dihydroxypyrimidine (**4AP**) as the dopant [31]. This modified PTI showed red-shifted absorption spectrum compared to the crystalline PTI and more similar to the melon structure, which improves the visible light absorption. By using this doped PTI, AQY of hydrogen evolution can be improved five times (3.4%), which is more than that of raw melon (0.6%). By supramolecular aggregation allowed by ionic melt polycondensation (IMP) of melamine and 2,4,6-triaminopyrimidine as dopants, Bhunia et al. introduced a new crystalline carbon nitride which exhibited an AQY of 15% for photocatalytic H_2 production at 400 nm [32]. Thus, molecular doping proved itself as very efficient technique for efficient H_2 production. Tri-*s*-triazine-based g-C_3N_4 is commonly prepared from the thermal polymerization of carbon and nitrogen containing organic and inorganic materials such as melamine, cyanamide, urea, thiourea, ammonium thiocyanate and guanidine sulphate [33–41]. These materials are known as efficient photocatalysts for splitting of water. The g-C_3N_4 polymers prepared from different starting materials usually show different properties such as surface morphology, optical absorption, charge carrier ability and electronic band structure. So, particularly by this way, the performance for photochemical reaction can be varied. To create porosity in the structure, strategies, such as hard template method, self-assembly (SA) and exfoliation, are conducted to improve the photocatalytic performances. Maeda et al. successfully introduced the spine nanosphere carbon nitride *in situ* deposited with 3 wt% Pt as co-catalysts as efficient photocatalyst for hydrogen generation with a hydrogen evolution rate (HER) of 574 μmol h^{-1} [42]. The AQY for H_2 evolution at 420 nm can reach a value of 9.6%.

3.4. Conjugated microporous polymers (CMPs)

Conjugated polymers with extended π-conjugation have been intensively explored in organic electronics and semiconductor device due to electronic and physical properties. Copolymerization of various monomeric units (building block) led to conjugated polymers, which allow to design various kind polymers with different electronic properties by varying the structural modification in monomeric units. Very recently, this kind of conjugated microporous polymers (CMPs) were successfully applied for photochemical hydrogen generation. Sprick et al. reported a series of CMP by varying the ratio of monomeric unit (**Figure 6**) [43]. The surface area is also tuneable

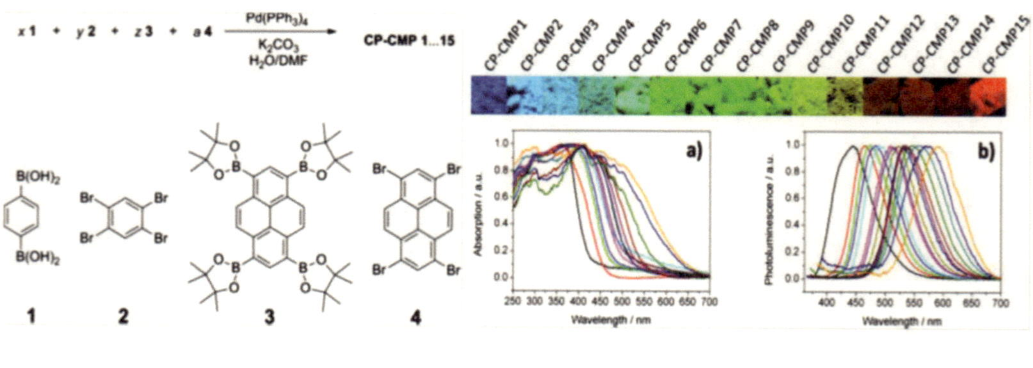

Figure 6. Schematic representation of reactions for the synthesis of CP-CMPs. Optical properties of the conjugated copolymer photocatalysts. Reproduced with permission from ACS; see Ref. [43].

by changing the monomers. Optical properties are also tuned by different amount of monomers which reflect in the range of optical band gap between 1.94 and 2.95 eV with an absorption band edge 445–588 nm. When in a water/methanol mix solvent in the presence of TEOA, photocatalytic reaction to generate hydrogen of these materials was tested. **CP-CMP10** shows the highest activity with a hydrogen evolution rate of 17.4 μmol h^{-1}. These polymers even showed hydrogen evolution performance without any co-catalyst such as Pt. It showed the optimal AQY for visible light H_2 production is 2.62% in the presence of 80 vol.% diethylamine as the sacrificial agent.

Subsequently, Li et al. introduced donor-acceptor type microporous polymer by co-polymerization method [44]. Two types of monomeric units, electron-rich units (**M2-M4**) and electron-deficient (**M1**) units, were copolymerized with different ration of weak donor biphenyl and weak acceptor bipyridine to produce a series of four polymers (**Figure 7**). The calculated band gaps of all polymers were in the range of 2.0–2.8 eV. The light absorbing ability of these polymers is significantly improved. Charge separation is facilitated due to the intra-molecular charge transfer, which is in fact very essential to generate excited holes and electrons for water oxidation and proton reduction, respectively. **PCP4e** showed the best performance for photocatalytic hydrogen production with a HER of 33.0 μmol h^{-1}. Bipyridyl-substituted polymers showed better performance than the biphenyl-containing polymer which can be attributed due to the presence of more electronegative N-atom, which may facilitate hydrogen adsorption through hydrogen bonding to enhance the proton reduction process.

On the other hand, Yang et al. reported benzothiadiazole-based donor-acceptor type conjugated microporous polymers as photocatalysts for hydrogen generation [15]. Benzothiadiazole unit is highly explored in organic conjugated materials for electronic devices due to its unique electron withdrawing ability. By co-polymerization technique starting with benzene and benzothiadiazole as basic monomeric unit and changing the substitution in benzene ring, they are able to synthesize a series of conjugated polymer from 1D to 3D in nature (**Figure 8**). **B-BT-1,4**

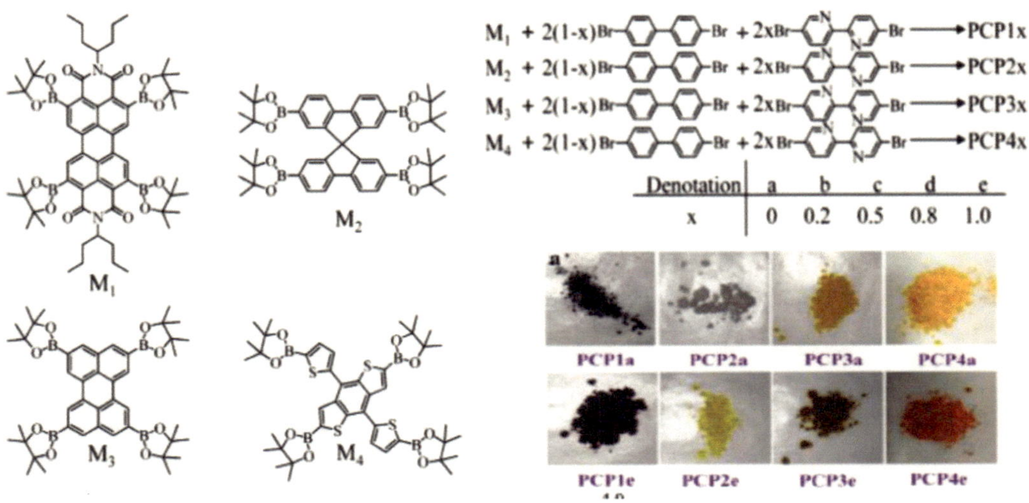

Figure 7. Structures of monomers and synthesis of PCP photocatalysts by Suzuki polycondensation and photographs of the prepared **PCPa** and **PCPe** photocatalysts. Reproduced with permission from ACS; see Ref. [44].

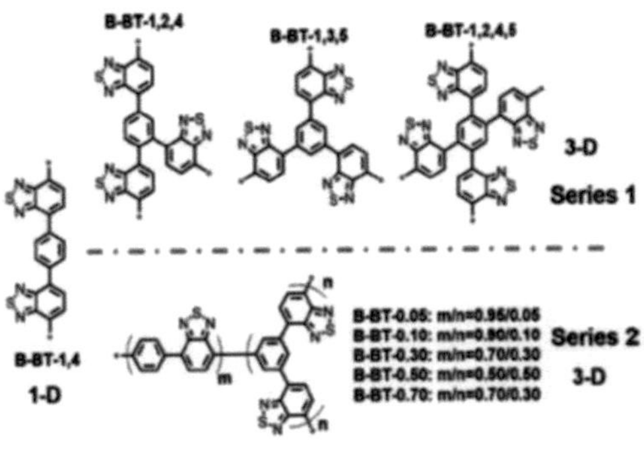

Polymer	S_{BET}[a] $[m^2 g^{-1}]$	PD[b] [nm]	PV[c] $[m^3 g^{-1}]$	HOMO[d]/ LUMO[e] [V vs. NHE]	OG[f] [eV]	HER[g] $[\mu mol\, h^{-1}]$
B-BT-1,4	39	0.14	4.0	1.28/−0.89	2.17	116
B-BT-1,2,4	17	0.03	4.5	1.27/−0.98	2.25	1.3
B-BT-1,3,5	280	0.37	1.5	1.41/−1.03	2.44	20
B-BT-1,2,4,5	40	0.06	1.5	1.23/−1.09	2.32	0.4
B-BT-0.05	58	0.10	3.8	1.25/−0.95	2.20	82
B-BT-0.10	93	0.24	3.8	1.24/−0.97	2.21	50
B-BT-0.30	129	0.51	4.5	1.23/−0.99	2.22	39
B-BT-0.50	40	0.20	10.7	1.23/−1.00	2.23	19
B-BT-0.70	84	0.20	3.8	1.27/−1.02	2.29	19

Figure 8. Structures of two series of polybenzothiadiazoles with different molecular designs and porosity data and electrochemical properties of the polymers. Reproduced with permission from Wiley-VCH; see Ref. [15].

itself showed photocatalytic activity with a HER of 12.0 $\mu mol\ h^{-1}$. Upon addition of 3 wt% Pt as cocatalyst, the HER was enhanced to 116.0 $\mu mol\ h^{-1}$. The AQY of **B-BT-1,4** for hydrogen evolution was found to be as high as 4.01%.

Chu et al. reported crystalline polyimide (PI) via high temperature dehydration reaction between melamine (MA) and pyromellitic dianhydride (PMDA) (**Figure 9**) [45]. A series of PIs with different optical band gaps (2.56–3.39 eV) were synthesized and the degree of polymerization is controlled by tuning the heating temperature. These polymers showed photocatalytic activity to produce hydrogen from water in the presence of Pt and TEOA as cocatalyst and sacrificial donor, respectively. Out of all PIs, **PI-325** showed the best activity with a HER of 150 $\mu mol\ h^{-1}$ and 15.2 $\mu mol\ h^{-1}$ under UV and visible light irradiation, respectively. **PI-325** showed AQY of 0.2% at 420 nm light irradiation.

In early 2010, Schwab et al. successfully introduced a series of imine-based conjugate poly (azomethine) materials [10]. The Schiff-base condensation reaction were adopted between the 1,3,5-tris(4- aminophenyl)benzene (**1**) as amine residue with different aldehydes such as 1,4-phthalaldehyde (**2**), naphthalene-2,6-dicarbaldehyde (**3**), anthracene-2,6-dicarbaldehyde (**4**) and anthracene-9,10-dicarbaldehyde (**5**) to synthesize the polymers. The optical band gaps of the synthesized polymers were located in the range of 1.96–2.38 eV with an optical band edge of 532–632 nm. These polymers showed photocatalytic activity under UV-light irradiation with a HER of 7 $\mu mol\ h^{-1}$ (**ANW-2**), whereas under visible light these polymers did not show any photocatalytic activity (**Figure 10**).

3.5. Conjugated linear and planarized polymers

As we mentioned before, Yanagida et al. in early 1985 reported poly(p-phenylene) as active photochemical hydrogen production catalyst under UV-light illumination [6]. Due to the solubility and low activity, there is no further advancement in this area. Recently, Sprick et al. synthesized and characterized a series of planar polymers by comprising phenylene and fluorine, carbazole, or dibenzo [b,d] thiophene [46]. From the study, they concluded that planar polymers showed higher photocatalytic activity to produce hydrogen. The pure copolymers

http://dx.doi.org/10.5772/intechopen.70444

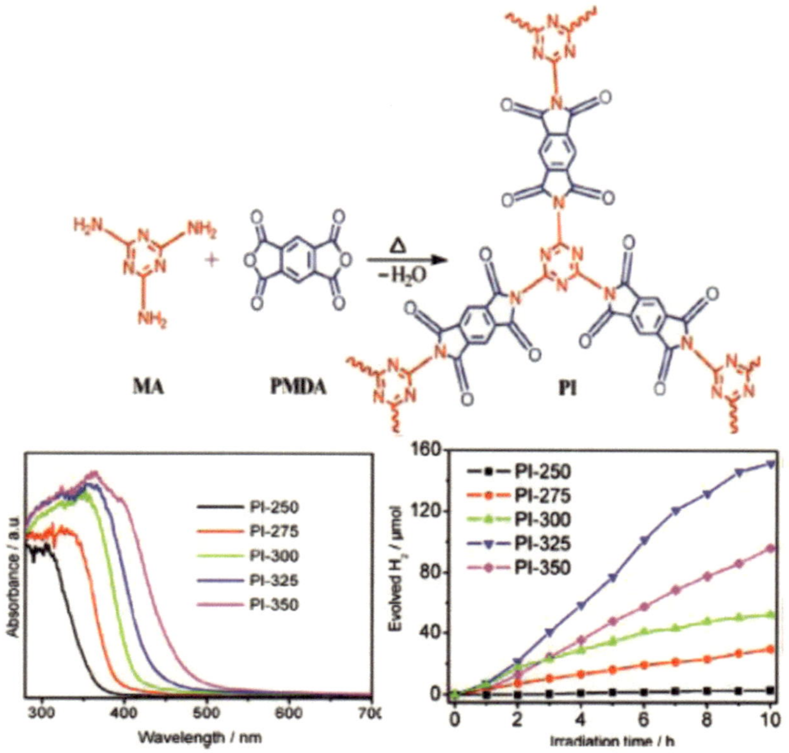

Figure 9. Reaction scheme for the synthesis of PI, UV-vis absorption spectra of PI samples processed at different temperatures and time course of H_2 evolution from a 10 vol% aqueous methanol solution by Pt-deposited PI samples under visible light irradiation (λ>420 nm). Reproduced with permission from Elsevier; see Ref. [45].

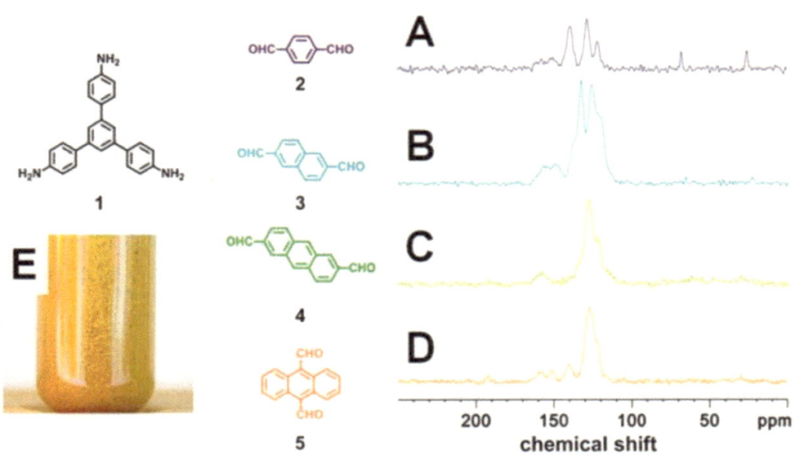

Figure 10. Molecular structures of the building units for the **ANW** materials and time course of H_2 production through **ANW** networks from water containing 10vol. % triethanolamine as electron donor (λ>300 nm). Reproduced with permission from RSC; see Ref. [10].

(**P1-P7**) are also exhibited excellent visible light-induced H_2 evolution activities even without depositing any co-catalysts, whereas methanol and triethylamine mixture were used as the sacrificial electron donors. Out of all planar co-polymers, **P7** showed the highest photocatalytic activity with HER of 92 μmol h^{-1}, while the rate was further found to be increased up to 116 μmol h^{-1} in the presence of Pt-nanoparticles as a co-catalyst. The AQY of hydrogen generation for **P7** was measured as 2.3% at 420 nm (**Figure 11**).

3.6. Polymer dots (Pdots)

Till discussed all materials behaved as heterogeneous catalyst because these all have poor dispersibility in water, to run the photocatalytic experiment in absolutely aqueous condition and the organic/water mixed solvent were commonly used to make good dispersibility. Another drawback is that majority materials were worked under UV light and also in the presence of a co-catalyst such as Pt. In the absence of co-catalyst and visible light, all show very unsatisfactory performances. Tian et al. used the conjugated polymer dots (Pdots), which is nicely dispersible in aqueous solution for photocatalytic proton reduction [47]. The Pdots of hydrophobic conjugated polymer (**PFBT**) were prepared through nanoprecipitation method by mixing with an amphiphilic co-polymers (**PS-PEG-COOH**). Typically, **PFBT** was dissolved in THF together with the co-polymer **PS-PEG-COOH** and then was poured into water. THF was removed later by continuously sonicating and Ar purging. The well-dispersed nanoparticles in water were obtained. Particle size of the nanoparticles was around 30–50 nm measured by dynamic light scattering (DLS). The optical band gap of PFBT Pdots was 2.38 eV. Without any co-catalyst in the presence of ascorbic acid as sacrificial donor, Pdots showed excellent photocatalytic activity with a HER of 8.3±0.2 mmol h^{-1} g^{-1}. Pdots configuration is very important because the pristine polymer itself did not show activity under the same test condition. The AQY of hydrogen evolution was 0.5% at 445 nm. However, after 1.5 h, the Pdots deactivated and precipitated out from the aqueous medium by forming bigger particle (**Figure 12**).

In order to further improve the system and study the reactivity sites of photocatalysis, Tian et al. subsequently investigated structures insight for highly efficient Pdots by both experimental and theoretical prospect. Pdots prepared from **1** and **2** (**Figure 13**) with benzothiadiazole (BT) units showed photocatalytic activity with a HER of 8.3±0.2 mmol h^{-1} g^{-1} and 50±0.5 mmol h^{-1} g^{-1}, whereas Pdots of **3** without such a unit did not show any photocatalytic activity. It suggests that the BT unit has crucial role in proton reduction reaction [48]. The theoretical study suggests that proton adsorption is facilitated due to the presence of N sites in the electron withdrawing BT unit which in fact enhances the photocatalytic activity. With

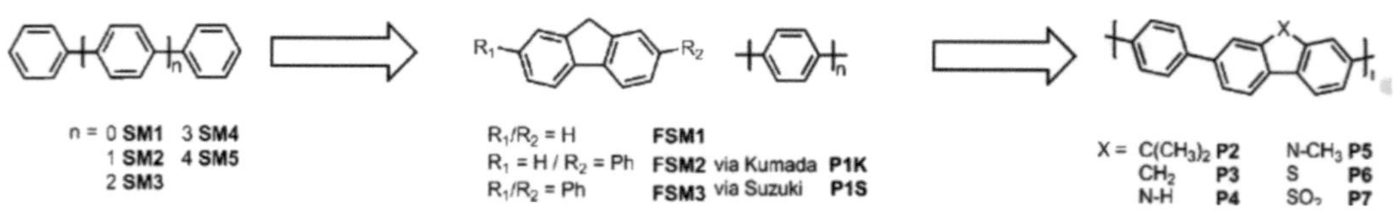

Figure 11. Schematic representation of reactions for the synthesis of planarized conjugated polymers. Reproduced with permission from Wiley-VCH; see Ref. [46].

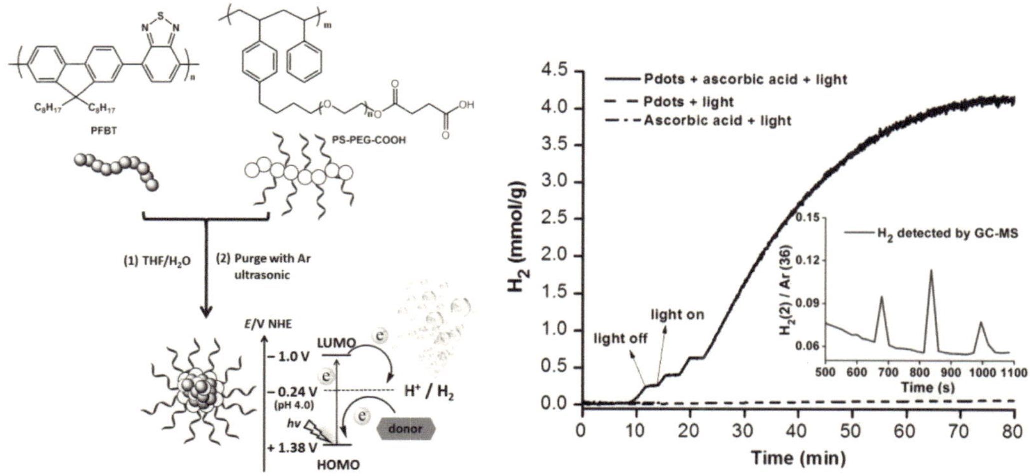

Figure 12. Preparation of PFBT Pdots and the diagram for light-driven hydrogen generation; visible light-driven hydrogen generation from water at room temperature. Conditions: **PFBT** Pdots, ca. 16.8 mg/mL; ascorbic acid: 0.2 M; pH 4.0 (adjusted with 1 M NaOH); LED lamp: white light (l>420 nm); the kinetic curve is monitored by a hydrogen sensor (see details in the Supporting Information). Inset: H_2 generation confirmed by GC-MS, the three peaks are signals from three injections of the headspace gas in the reaction flask. Reproduced with permission from Wiley-VCH; see Ref. [47].

respect to **Pdots 1**, **Pdots 2** shows even more stability and works for more than 4 hour. The AQY for H_2 generation of **Pdots 2** was measured as 0.6% at 550 nm. The isotopic labelling experiment, using D_2O and NaOD instead of H_2O and NaOH in a photocatalytic experiment of **Pdots 2**, suggests that the evolving hydrogen is mainly coming from the water. Since this kind of conjugated polymers is prepared by Pd-catalysed copolymerization method, where the residual Pd is known to be a good catalyst for proton reduction and it could behave as co-catalyst for this photocatalytic reaction. However, from the Pd poisoning experiments with carbon monoxide (CO) and ethylenediaminetetraacetic acid (EDTA), it suggests that the Pdots

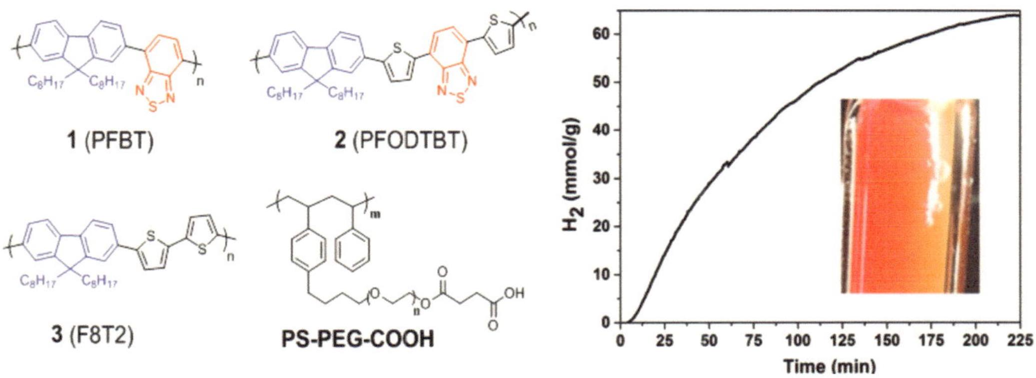

Figure 13. Molecular structures of three organic D–A polymers (**1**, **2** and **3**) and a co-polymer PS-PEG-COOH; light-driven hydrogen generation of Pdots **1** (17 μg/ml), **2** (13 μg/ml) and **3** (19 μg/ml) in water at ambient temperature. Conditions: ascorbic acid: 0.2 M; pH 4.0 (adjusted with 1 M NaOH); white LED light (420 nm, 17 W, 5000 K). Inset: H_2 bubbles observed during photocatalysis. Reproduced with permission from RSC; see Ref. [48].

are the real photocatalysts. The DFT computation also suggests the unique structure of the Pdots from aggregated polymers benefits for lowering the energy barrier of proton reduction process via N atoms in BT units.

4. Perspective

Library of carbon-based organic photocatalysts are very few in comparison to the inorganic photocatalysts. The performance of these materials are still unsatisfactory but these organic materials have several advantages such as light-weight, easily processable and better reproducibility upon batch-to-batch variation. The organic photocatalysts are of great interest because there is a wide opportunity to introduce several new materials with different optical and electrochemical properties by small structural modification. By proper structural modification, the performance of these materials can be tuned and would be comparable with the inorganic materials. The major drawback of all of these organic carbon-based material is water dispersibility because of very poor solubility in water. The polymer dot encounters the solubility issue by some extent whereas rather need optimization to solve the associated problem such as unwanted aggregation along with light illumination, which causes deactivation and so on.

Acknowledgements

The authors thanks the financial support from KA Wallenberg Foundation, Swedish Energy Agency, Sweden VR-Korea NRF joint climate change project, Stiftelsen Olle Engkvist Byggmästare and Göran Gustafssons Stiftelse.

Author details

Palas Baran Pati[1], SU-IL In[2] and Haining Tian[1]*

*Address all correspondence to: haining.tian@kemi.uu.se

1 Physical Chemistry, Department of Chemistry-Ångström Laboratory, Uppsala University, Uppsala, Sweden

2 Department of Energy Systems Engineering, DGIST, Daegu, Republic of Korea

References

[1] Hagfeldt A, Boschloo G, Sun L, Kloo L, Pettersson H. Chemical Reviews. Dye-sensitized solar cells. 2016;**110**:6595

[2] Liu C, Wang K, Gong X, Heeger AJ. Chemical Society Reviews. Low bandgap semiconducting polymers for polymeric photovoltaics. 2016;**45**:4825

[3] Kudo A, Miseki Y. Chemical Society Reviews. Heterogeneous photocatalyst materials for watersplitting. 2009;**38**:253

[4] Fujishima A, Honda K. Nature. Electrochemical photolysis of water at a semiconductor electrode. 1972;**238**:37

[5] Ran J, Zhang J, Yu J, Jaroniecc M, Qiao SZ. Chemical Society Reviews. Earth-abundant cocatalysts for semiconductor-based photocatalytic water splitting. 2014;**43**:7787

[6] Yanagida S, Kabumoto A, Mizumoto K, Pac C, Yoshino K. Journal of the Chemical Society, Chemical Communications. Poly(p-phenylene)-catalysed photoreduction of water to hydrogen. 1985;**8**:474

[7] Wang X, Maeda K, Thomas A, Takanabe K, Xin G, Carlsson JM, Domen K, Antonietti M. Nature Materials. A metal-free polymeric photocatalyst for hydrogen production from water under visible light. 2009;**8**:76

[8] Ye S, Wang R, Wu M-Z, Yuan Y-P. Applied Surface Science. A review on g-C3N4 for photocatalytic water splitting and CO_2 reduction. 2015;**358**:15

[9] Liu J, Liu Y, Liu N, Han Y, Zhang X, Huang H, Lifshitz Y, Lee S-T, Zhong J, Kang Z. Science. Water splitting. Metal-free efficient photocatalyst for stable visible water splitting via a two-electron pathway. 2015;**347**:970-974

[10] Schwab MG, Hamburger M, Feng X, Shu J, Spiess HW, Wang X, Antonietti M, Mullen K. Chemical Communications. Photocatalytic hydrogen evolution through fully conjugated poly(azomethine) networks. 2010;**46**:8932

[11] Stegbauer L, Schwinghammer K, Lotsch BV. Chemical Science. A hydrazone-based covalent organic framework for photocatalytic hydrogen production. 2014;**5**:2789

[12] Schwinghammer K, Hug S, Mesch MB, Senker J, Lotsch BV. Energy & Environmental Science. Phenyl-triazine oligomers for light-driven hydrogen evolution. 2015;**8**:3345

[13] Park JH, Ko KC, Park N, Shin H-W, Kim E, Kang N, Hong JK, Lee SM, Kim HJ, Ahn TK, Lee JY, Son SU. Journal of Materials Chemistry A. Microporous organic nanorods with electronic push–pull skeletons for visible light-induced hydrogen evolution from water. 2014;**2**:7656

[14] Kailasam K, Schmidt J, Bildirir H, Zhang G, Blechert S, Wang X, Thomas A. Macromolecular Rapid Communications. Room temperature synthesis of heptazine-based microporous polymer networks as photocatalysts for hydrogen evolution. 2013;**34**:1008

[15] Yang C, Ma BC, Zhang L, Lin S, Ghasimi S, Landfester K, Zhang KAI, Wang X. Angewandte Chemie International Edition. Molecular engineering of conjugated polybenzothiadiazoles for enhanced hydrogen production by photosynthesis. 2016;**55**:9202

[16] Simon T, Bouchonville N, Berr MJ, Vaneski A, Adrović A, Volbers D, Wyrwich R, Döblinger M, Susha AS, Rogach AL, Jäckel F, Stolarczyk JK, Feldmann J, Nature Materials. Redox shuttle mechanism enhances photocatalytic H_2 generation on Ni-decorated CdS nanorods. 2014;**13**:1013

[17] Wang ZJ, Ghasimi S, Landfester K, Zhang KAI. Advanced Materials. Molecular structural design of conjugated microporous poly(benzooxadiazole) networks for enhanced photocatalytic activity with visible light. 2015;**27**:6265

[18] Zhang K, Kopetzki D, Seeberger PH, Antonietti M, Vilela F. Angewandte Chemie International Edition. Surface area control and photocatalytic activity of conjugated microporous poly(benzothiadiazole) networks. 2013;**52**:1432

[19] Vyas VS, Haase F, Stegbauer L, Savasci G, Podjaski F, Ochsenfeld C, Lotsch BV. Nature Communications. A tunable azine covalent organic framework platform for visible light-induced hydrogen generation. 2015;**6**:8508

[20] Haase F, Banerjee T, Savasci G, Ochsenfeld C, Lotsch BV. Faraday Discussions. Structure–property–activity relationships in a pyridine containing azine-linked covalent organic framework for photocatalytic hydrogen evolution. DOI: 10.1039/C7FD00051K

[21] Kuhn P, Antonietti M, Thomas A. Angewandte Chemie International Edition. Porous, covalent triazine-based frameworks prepared by ionothermal synthesis. 2008;**47**:3450

[22] Wang XC, Maeda K, Chen XF, Takanabe K, Domen K, Hou YD, Fu XZ, Antonietti M. Journal of the American Chemical Society. Polymer semiconductors for artificial photosynthesis: hydrogen evolution by mesoporous graphitic carbon nitride with visible light. 2009;**131**:1680

[23] Zheng Y, Lin LH, Ye XJ, Guo FS, Wang XC. Angewandte Chemie International Edition. Helical graphitic carbon nitrides with photocatalytic and optical activities. 2014;**53**:11926

[24] Jun YS, Park J, Lee SU, Thomas A, Hong WH, Stucky GD. Angewandte Chemie International Edition. Three-dimensional macroscopic assemblies of low-dimensional carbon nitrides for enhanced hydrogen evolution. 2013;**52**:11083

[25] Zheng Y, Lin LH, Wang B, Wang XC. Angewandte Chemie International Edition. Graphitic carbon nitride polymers toward sustainable photoredox catalysis. 2015;**54**:12868

[26] Zhao Y, Zhao F, Wang X, Xu C, Zhang Z, Shi G, Qu L. Angewandte Chemie International Edition. Graphitic carbon nitride nanoribbons: graphene-assisted formation and synergic function for highly efficient hydrogen evolution. 2014;**53**:13934

[27] Zhang MW, Wang XC. Energy & Environmental Science. Two dimensional conjugated polymers with enhanced optical absorption and charge separation for photocatalytic hydrogen evolution. 2014;**7**:1902

[28] Zhang GG, Wang XC. Journal of Catalysis. A facile synthesis of covalent carbon nitride photocatalysts by co-polymerization of urea and phenylurea for hydrogen evolution. 2013;**307**:246

[29] Han Q, Wang B, Zhao Y, Hu C, Qu L. Angewandte Chemie International Edition. A graphitic-C_3N_4 "seaweed" architecture for enhanced hydrogen evolution. 2015;**54**:11433

[30] Ham Y; Maeda K, Cha D, Takanabe K, Domen K. Chemistry, An Asian Journal. Synthesis and photocatalytic activity of poly(triazine imide). 2013;**8**:218

[31] Schwinghammer K, Tuffy B, Mesch MB, Wirnhier E, Martineau C, Taulelle F, Schnick W, Senker J, Lotsch BV. Angewandte Chemie International Edition. Triazine-based carbon nitrides for visible-light-driven hydrogen evolution. 2013;**52**:2435

[32] Bhunia MK, Yamauchi K, Takanabe K. Angewandte Chemie International Edition. Harvesting solar light with crystalline carbon nitrides for efficient photocatalytic hydrogen evolution. 2014;**53**:11001

[33] Kuriki R, Sekizawa K, Ishitani O, Maeda K. Angewandte Chemie International Edition. Visible-light-driven CO_2 reduction with carbon nitride: enhancing the activity of ruthenium catalysts. 2015;**54**:2406

[34] Cui YJ, Ding ZX, Fu XZ, Wang XC. Angewandte Chemie International Edition, Construction of conjugated carbon nitride nanoarchitectures in solution at low temperatures for photoredox catalysis. 2012;**51**:11814

[35] Shalom M, Inal S, Fettkenhauer C, Neher D, Antonietti M. Journal of the American Chemical Society. Improving carbon nitride photocatalysis by supramolecular preorganization of monomers. 2013;**135**:7118

[36] Zheng Y, Liu J, Liang J, Jaroniecc M, Qiao SZ. Energy & Environmental Science. Graphitic carbon nitride materials: controllable synthesis and applications in fuel cells and photocatalysis. 2012;**5**:6717

[37] Martin J, Qiu KP, Shevlin SA, Handoko AD, Chen XW, Guo ZX, Tang JW. Angewandte Chemie International Edition. Highly efficient photocatalytic H2 evolution from water using visible light and structure-controlled graphitic carbon nitride. 2014;**53**:9240

[38] Zhang GG, Lan ZA, Lin LH, Lin S, Wang XC. Chemical Science. Overall water splitting by Pt/g-C_3N_4 photocatalysts without using sacrificial agents. 2016;**7**:3062

[39] Zhang GG, Zhang JS, Zhang MW, Wang XC. Journal of Materials Chemistry. Polycondensation of thiourea into carbon nitride semiconductors as visible light photocatalysts. 2012;**22**:8083

[40] Cui YJ, Zhang JS, Zhang GG, Huang JH, Liu P, Antonietti M, Wang XC. Journal of Materials Chemistry. Synthesis of bulk and nanoporous carbon nitride polymers from ammonium thiocyanate for photocatalytic hydrogen evolution. 2011;**21**:13032

[41] Long H, Lin JL, Wang XC. Journal of Materials Chemistry A. Thermally-induced desulfurization and conversion of guanidine thiocyanate into graphitic carbon nitride catalysts for hydrogen photosynthesis. 2014;**2**:2942

[42] Maeda K, Wang XC, Nishihara Y, Lu DL, Antonietti M, Domen K. Journal of Physical Chemistry C. Photocatalytic activities of graphitic carbon nitride powder for water reduction and oxidation under visible light. 2009;**113**:4940

[43] Sprick RS, Jiang JX, Bonillo B, Ren S, Ratvijitvech T, Guiglion P, Zwijnenburg MA, Adams DJ, Cooper AI. Journal of the American Chemical Society. Tunable organic photocatalysts for visible-light-driven hydrogen evolution. 2015;**137**:3265

[44] Li LW, Cai ZX, Wu QH, Lo WY, Zhang N, Chen LX, Yu LP. Journal of the American Chemical Society. Rational design of porous conjugated polymers and roles of residual palladium for photocatalytic hydrogen production. 2016;**138**:7681

[45] Chu S, Wang Y, Wang CC, Yang JC, Zou ZG. International Journal of Hydrogen Energy. Bandgap modulation of polyimide photocatalyst for optimum H2 production activity under visible light irradiation. 2013;**38**:10768

[46] Sprick RS, Bonillo B, Clowes R, Guiglion P, Brownbill NJ, Slater BJ, Blanc F, Zwijnenburg MA, Adams DJ, Cooper AI. Angewandte Chemie International Edition. Visible-light-driven hydrogen evolution using planarized conjugated polymer photocatalysts. 2016;**55**:1792

[47] Wang L, Fernández-Terán R, Zhang L, Fernandes DLA, Tian L, Chen H, Tian H. Angewandte Chemie International Edition. Organic polymer dots as photocatalysts for visible light-driven hydrogen generation. 2016;**55**:12306-12310

[48] Pati PB, Damas G, Tian L, Fernandes DLA, Zhang L, Pehlivan IB, Edvinsson T, Araujo CM, Tian H. Energy & Environmental Science. An experimental and theoretical study of an efficient polymer nano-photocatalyst for hydrogen evolution. 2017;**10**:1372